Avidin–Biotin Interactions

METHODS IN MOLECULAR BIOLOGY™

John M. Walker, SERIES EDITOR

METHODS IN MOLECULAR BIOLOGY™

Avidin–Biotin Interactions

Methods and Applications

Edited by

Robert J. McMahon

Mead Johnson Nutritionals, Evansville, IN

 Humana Press

Editor
Robert J. McMahon
Mead Johnson Nutritionals, Evansville, IN

Series Editor
John M. Walker, Professor Emeritus
School of Life Sciences,
University of Hertfordshire
Hatfield
Hertfordshire AL10 9AB, UK.

ISBN: 978-1-58829-583-5   e-ISBN: 978-1-59745-579-4

Library of Congress Control Number: 2007933142

Cover illustration: Triple-labelled confocal microscope image of human cervical carcinoma cell (HeLa cell) after labelling with Molecular Probes' MitoTracker Red CMXRos (red fluorescence) to visualize mitochondria, anti-biotin antibody labelled with Alexa Fluor®-488 secondary antibody (green fluorescence) to identify endogenous biotin and DAPI (blue fluorescence) to counterstain the nuclei. The presence of endogenous biotin in mitochondria results in co-localisation of red and green signals producing yellow mitochondria. (Courtesy of Elisabeth Coene and the BioImaging Facility in the Dunn School of Pathology.)

Printed on acid-free paper

9 8 7 6 5 4 3 2 1

springer.com

Preface

For those who labor in the biomedical sciences, there can be no denying that the proliferation of readily available and relatively low cost, prefabricated kits has played a significant and positive role in moving many fields of research forward. These kits often reduce the steep learning curve-associated new methodology, remove uncertainty, and allow many researchers to move into areas that previously would have been completely off limits due to the large start up investment that modern biological methodology requires. Despite the opportunities that such prefabricated kits provide to both the student and the experienced researcher, it does not come without some cost. Inherent in these kits is the need to adhere to strictly determined procedures, as deviation from these procedures often leads to lower quality and less reliable results. Additionally, because many of the modern kits incorporate reagents that are the valuable intellectual property of the company that designed them, it is often difficult for the student or the researcher to fully understand the required chemistry underlying the methodology. This lack of clarity limits the potential usefulness of any kit, as although the kit is extremely well designed to perform under a specific set of known conditions, the very nature of biomedical research is to push into unknown areas using current methodology modified to tackle the new frontier.

This book is specifically about the application of one particular methodology, that which exploits the extremely powerful interaction between the protein avidin or its homolog and the vitamin biotin and some of its homolog. Because the molecular interaction of these two entities exhibits such high affinity, tagging of one or both parts of this pair allows sensitive and specific detection of biomolecules. For instance, if you have an antibody against a particular protein, you can "tag" this antibody with biotin, and subsequent binding by avidin conjugated to a reporter enzyme will allow detection and quantification. Many variations on this simple theme have since been developed, and today the avidin–biotin interaction is one of the most widely exploited in the biomedical sciences.

In spite of this wide use, the avidin–biotin methodology is well suited to the theme of the problem with kits. Although avidin–biotin-based kits have been produced for quite some time and remains a widely used methodology, many of the chapters in this text describe situations in which the previous methodology failed due to some factor, and the authors developed new and better ways for the avidin–biotin system to be utilized. Take for instance the example of the chapter by Willem Kamphuis and Jan Klooster, these authors describe that while detection systems based upon avidin–biotin interactions had been used in tissue sections for some time, there has historically been a problem with "non-specific" staining which confounded interpretation of results. With more careful study, it was eventually found that the "non-specific" detection by the avidin–biotin system was at least in part due to the detection of true endogenous biotin where it was not previously known to exist. Subsequently, strategies were developed to deal with the endogenous biotin in tissue sections that improved the utility of the basic technique.

In the end, this is perhaps the most valuable part of a text such as this; for the student or the researcher who wishes to begin or refresh their use of the avidin–biotin methodology, look at each of the enclosed chapters as not only excellent descriptions of protocols that one can use in the lab, but also as the expert learnings of researchers who were using a specific methodology, found a problem with the methodology in a new setting, and then devised ways to minimize or obviate the limitation of the technology. Applying this sort of investigative approach to the use of any methodology is critical to the advancement of science.

Robert J. McMahon

Contents

Contributors

SABINE ANDRE • *Charles University, 1st Faculty of Medicine, Institute of Anatomy, U Nemocnice, Czech Republic*

MAHESH K. BHALGAT • *Molecular Probes, Inc., Eugene, OR*

ANNA BOGUSIEWICZ • *Department of Biochemistry and Molecular Biology, Department of Pediatrics, University of Arkansas for Medical sciences, Little Rock, Arkansas*

YAP CHING CHEW • *Department of Nutrition and Health Sciences, University of Nebraska at Lincoln, Lincoln, NE*

ELIZABETH D. COENE • *Sir Willam Dunn School of Pathology, University of Oxford, UK*

ANDREAS EBNER • *University of Linz, Linz, Austria*

GIULIANO ELIA • *Institute of Pharmaceutical Sciences, Swiss. Federal Institute of Technology, Zurich, Switzerland*

RUTH FREITAG • *University of Bayreuth, Bayreuth, Germany*

HERMANN J. GRUBER • *University of Linz, Linz, Austria*

SHALINI GUPTA • *University of Chicago Research Resources Center, Chicago, IL*

CHRISTOPH D. HAHN • *University of Linz, Linz, Austria*

ROSARIA P. HAUGLAND • *Molecular Probes, Inc., Eugene, OR*

FRANK HILBRIG • *University of Bayreuth, Bayreuth, Germany*

NICOLAS HUMBERT • *Institute of Chemistry, University of Neuchatel, Neuchatel, Switzerland*

ALBERT JELTSCH • *School of Engineering and Science, International University Bremen, Bremen, Germany*

GERALD KADA • *University of Linz, Linz, Austria*

KARL KAISER • *University of Linz, Linz, Austria*

WILLEM KAMPHUIS • *Netherlands Opthalmic Research Institute (NORI-KNAW), Department of Opthalmogenetics, Glaucoma Research Group, Meibergdreef 47, 1105 BA Amsterdam, The Netherlands*

JAN KLOOSTER • *Netherlands Opthalmic Research Institute (NORI-KNAW), Department of Opthalmogenetics, Glaucoma Research Group, Meibergdreef 47, 1105 BA Amsterdam, The Netherlands*

SANGEETH KRISHNANCHETTIAR • *University of Chicago Research Resources Center, Chicago, IL*

ALICE KUEH • *Department of Nutrition and Health Sciences, University of Nebraska at Lincoln, Lincoln, NE*

BERND LACKNER • *University of Linz, Linz, Austria*

SYED SALMAN LATEEF • *University of Chicago Research Resources Center, Chicago, IL*

BAO-SHIANG LEE • *University of Chicago Research Resources Center, Chicago, IL*

KIRSTEN LIEBERT • *School of Engineering and Science, International University Bremen, Bremen, Germany*

MARKUS MAREK • *University of Linz, Linz, Austria*

KRISTA MCCUTCHEON • *Antibody Engineering Department, Genentech Inc., South San Francisco, CA*

BRUCE E. MCKAY • *Hotchkiss Brain Institute, University of Calgary, Calgary, Alberta, Canada*

DONALD M. MOCK, M.D. • *Department of Biochemistry and Molecular Biology, Department of Pediatrics, University of Arkansas for Medical sciences, Little Rock, Arkansas*

MICHAEL L. MOLINEUX • *Hotchkiss Brain Institute, University of Calgary, Calgary, Alberta, Canada*

DARIO NERI • *Institute of Pharmaceutical Sciences, Swiss. Federal Institute of Technology, Zurich, Switzerland*

JEAN-MARC NEUHAUS • *Institute of Chemistry, University of Neuchatel, Neuchatel, Switzerland*

CHRISTOPH ROSLI • *Institute of Pharmaceutical Sciences, Swiss. Federal Institute of Technology, Zurich, Switzerland*

JASCHA-N. RYBAK • *Institute of Pharmaceutical Sciences, Swiss. Federal Institute of Technology, Zurich, Switzerland*

GAUTAM SARATH • *Department of Nutrition and Health Sciences, University of Nebraska at Lincoln, Lincoln, NE*

PETER SCHURMANN • *Institute of Chemistry, University of Neuchatel, Neuchatel, Switzerland*

MICHAEL K. SHAW • *Sir Willam Dunn School of Pathology, University of Oxford, UK*

KAREL SMETANA, JR. • *Charles University, 1st Faculty of Medicine, Institute of Anatomy, U Nemocnice, Czech Republic*

DAVID J. VAUX • *Sir Willam Dunn School of Pathology, University of Oxford, UK*

THOMAS R. WARD • *Institute of Chemistry, University of Neuchatel, Neuchatel, Switzerland*

WENDY W. YOU • *Department of Biochemistry and Biophysics, Oregon State University, Corvallis, OR*

JANOS ZEMPLENI • *Department of Nutrition and Health Sciences, University of Nebraska at Lincoln, Lincoln, NE*

ANDREA ZOCCHI • *Institute of Chemistry, University of Neuchatel, Neuchatel, Switzerland*

1

Preparation of Avidin Conjugates

Rosaria P. Haugland and Mahesh K. Bhalgat

1. Introduction

The high-affinity avidin–biotin system has found applications in different fields of biotechnology, including immunoassays, histochemistry, affinity chromatography, and drug delivery, to name a few. A brief description of avidin and avidin-like molecules, streptavidin, deglycosylated avidin, and NeutraLite avidin is presented in the Chapter 2. With four biotin-binding sites per molecule, the avidin family of proteins is capable of forming tight complexes with one or more biotinylated compounds (1). Typically, the avidin–biotin system is used to prepare signal-amplifying "sandwich" complexes between specificity reagents (e.g., antibodies) and detection reagents (e.g., fluorophores, enzymes, and so on). The specificity and detection reagents are independently conjugated, one with avidin and the other with biotin, or both with biotin, providing synthetic flexibility (2).

Avidin conjugates of a wide range of fluorophores, phycobiliproteins, secondary antibodies, microspheres, ferritin, and enzymes commonly used in immunochemistry are available at reasonable prices, making their small-scale preparation impractical and not cost effective (see **Note 1**). However, conjugations of avidin to specific antibodies, to uncommon enzymes, and to other proteins and peptides are often performed on-site. A general protocol for the conjugation of avidin to enzymes, antibodies, and other proteins is described in this chapter.

Avidin conjugates of oligodeoxynucleotides are hybrid molecules that not only provide multiple biotin-binding sites, but can also be targeted to complimentary DNA or RNA sequences, by annealing interactions. Such conjugates are useful for the construction of macromolecular assemblies with a wide

From: *Methods in Molecular Biology, vol. 418: Avidin-Biotin Interactions, Methods and Applications*
Edited by: R. J. McMahon © Humana Press, Totowa, NJ

variety of constituents *(3)*. The protocol outlined in **Subheading 3.1.** can be modified (*see* **Note 2**) for the conjugation of oligonucleotides to avidin.

Streptavidin conjugates are also being evaluated for use in drug delivery systems. A two-step imaging and treatment protocol has been developed that involves injection of a suitably prepared tumor-specific monoclonal antibody, followed by a second reagent that carries an imaging or therapeutic agent, capable of binding to the tumor-targeted antibody *(4)*. Owing to complications associated with the injection of radiolabeled biotin *(5)*, conjugation of the imaging or therapeutic agent to streptavidin is being considered instead. A protocol for radioiodination of streptavidin using IODO-BEADS *(6)* is described in **Subheading 3.2.** Some other methods that have been developed include the iodogen method *(7,8)*, the Bolton–Hunter reagent method *(9)*, and a few that do not involve direct iodination of tyrosine residues *(10–13)*. Streptavidin–drug conjugates are also candidates for therapeutic agents. Synthesis of a streptavidin–drug conjugate involves making a chemically reactive form of the drug followed by its conjugation to streptavidin. The synthetic methodology thus depends on the structure of the specific drug to be conjugated *(14–16)*.

The avidin–biotin interaction can also be exploited for affinity chromatography; however, there are limitations to this application. For example, a biotinylated protein captured on an avidin affinity matrix would likely be denatured by the severe conditions required to separate the high-affinity avidin–biotin complex. On the other hand, an avidin affinity matrix may find utility in the removal of undesired biotinylated moieties from a mixture or for the purification of compounds derivatized with 2-iminobiotin. The biotin derivative 2-iminobiotin has reduced affinity for avidin, and its moderate binding to avidin at pH 9.0 is greatly diminished at pH 4.5 *(17)*. Another approach to reducing the affinity of the interaction is to denature avidin to its monomeric subunits. The monomeric subunits have greatly reduced affinity for biotin *(18)*. We describe here a protocol for preparing native *(19)* and monomeric avidin matrices *(20)*. Recently, modified streptavidins, hybrids of native and engineered subunits with lower binding constants, have been prepared that may also be suitable for affinity matrices *(21)*.

2. Materials

2.1. Conjugation with Antibodies and Enzymes

1. Avidin.
2. Antibody, enzyme, peptide, protein, or thiolated oligonucleotide to be conjugated to avidin.
3. Succinimidyl 3-(2-pyridyldithio)propionate (SPDP) (*see* **Note 3**).

4. Succinimidyl *trans*-4-(*N*-maleimidylmethyl)cyclohexane-1-carboxylate (SMCC).
5. Dithiothreitol (DTT).
6. Tris (2-carboxyethyl)phosphine (TCEP).
7. N-ethylmaleimide (NEM).
8. Anhydrous dimethyl sulfoxide (DMSO) or anhydrous dimethylformamide.
9. Phosphate buffer, 0.1 M: 0.1 M sodium phosphate, 0.1 M NaCl at pH 7.5. Dissolve 92 g of Na_2HPO_4, 21 g of $NaH_2PO_4 \cdot H_2O$, and 46.7 g of NaCl in approximately 3.5 L of distilled water and adjust the pH to 7.5 with 5 M NaOH. Dilute to 8 L. Store refrigerated.
10. Sodium bicarbonate, 1 M (*see* **Note 4**): Dissolve 8.4 g in 90 mL of distilled water and adjust the volume to 100 mL. A freshly prepared solution has a pH of 8.3–8.5.
11. Molecular exclusion matrix with properties suitable for purification of the specific conjugate. Sephadex G-200 (Pharmacia Biotech, Uppsala, Sweden), Bio-Gel A-0.5 m or Bio-Gel A-1.5 m (Bio-Rad Laboratories, Hercules, CA, USA) are useful for relatively small to large conjugates, respectively.
12. Sephadex G-25 (Pharmacia Biotech) or other equivalent matrix.

2.2. Radioiodination Using IODO-BEADS

1. Streptavidin.
2. $Na^{131}I$ or $Na^{125}I$, as desired.
3. IODO-BEADS (Pierce Chemical, Rockford, IL, USA).
4. Phosphate-buffered saline (PBS), pH 7.2: Dissolve 1.19 g of K_2HPO_4, 0.43 g of KH_2PO_4, and 9 g of NaCl in 900 mL of distilled water. Adjust the pH to 7.2 and dilute to 1 L with distilled water.
5. Saline solution: 9 g of NaCl dissolved in 1 L of distilled water.
6. Bovine serum albumin (BSA) (0.1%) solution in saline: 0.1 g of BSA dissolved in 100 mL of saline solution.
7. Trichloroacetic acid (TCA), 10% (w/v) solution in saline: Dissolve 1 g TCA in 10 mL of saline solution.
8. Bio-Gel P-6DG Gel (Bio-Rad).

2.3. Avidin Affinity Matrix

1. 50–100 mg of avidin.
2. Sodium borohydride.
3. 1,4-Butanediol-diglycidylether.
4. Succinic anhydride.
5. 6 M Guanidine HCl, in 0.2 M KCl/HCl, pH 1.5: Dissolve 1.5 g of KCl in 50 mL of distilled water. Add 57.3 g of guanidine HCl with stirring. Adjust the pH to 1.5 with 1 M HCl. Adjust the volume to 100 mL with distilled water.
6. 0.2 M Glycine-HCl, pH 2.0: Dissolve 22.3 g of glycine-HCl in 900 mL of distilled water. Adjust the pH to 2.0 with 6 M HCl and the volume to 1 L with distilled water.

7. PBS: *see* **Subheading 2.2., item 4**.
8. 0.2 M Sodium carbonate, pH 9.5: Dissolve 1.7 g of sodium bicarbonate in 80 mL of distilled water. Adjust the pH to 9.5 with 1 M NaOH and the volume to 100 mL with distilled water.
9. 0.2 M Sodium phosphate, pH 7.5: Weigh 12 g of Na_2HPO_4 and 2.5 g of $NaH_2PO_4.H_2O$ and dissolve in 900 mL of distilled water. Adjust the pH to 7.5 with 5 M NaOH and the volume to 1 L with distilled water.
10. 20 mM Sodium phosphate, 0.5 M NaCl, 0.02% sodium azide, pH 7.5: Dilute 100 mL of the buffer described in **item 9** (above) to 900 mL with distilled water. Add 28 g of NaCl and 200 mg of sodium azide. Adjust pH if necessary and dilute to 1 L with distilled water.
11. Sepharose 6B (Pharmacia Biotech) or other 6% crosslinked agarose gel.

3. Methods

3.1. Conjugation with Antibodies and Enzymes

3.1.1. Avidin Thiolation

An easy-to-use, protein-to-protein crosslinking kit is now commercially available (Molecular Probes, Eugene, OR, USA). This kit allows predominantly 1:1 conjugate formation between two proteins (0.2–3.0 mg) through the formation of a stable thioether bond *(22)*, with minimal generation of aggregates. A similar protocol is described here for conjugation of 5 mg avidin to antibodies or enzymes. Modifications of the procedure for conjugation of avidin to thiolated oligonucleotides and peptides are described in **Notes 2** and **5**, respectively. Although the protocol described in this section uses avidin for conjugation, it can be applied for the preparation of conjugates using either avidin, streptavidin, deglycosylated avidin, or NeutraLite avidin.

1. Dissolve 5 mg of avidin (76 nmol) in 0.5 mL of 0.1 M phosphate buffer to obtain a concentration of 10 mg/mL.
2. Weigh 3 mg of SPDP and dissolve in 0.3 mL of DMSO to obtain a 10 mg/mL solution. This solution must be prepared fresh immediately before using. Vortex or sonicate to ensure that the reagent is completely dissolved.
3. Slowly add 12 µL (380 nmol) of the SPDP solution (*see* **Note 3**) to the stirred solution of avidin. Stir for 1 h at room temperature.
4. Purify the thiolated avidin on a 7 × 250 mm size exclusion column, such as Sephadex G-25 equilibrated in 0.1 M phosphate buffer.
5. Determine the degree of thiolation (optional):

   a. Prepare a 100 mM solution of DTT by dissolving 7.7 mg of the reagent in 0.5 mL of distilled water.
   b. Transfer the equivalent of 0.3–0.4 mg of thiolated avidin (absorbance at 280 nm of a 1.0 mg/mL avidin solution = 1.54) and dilute to 1.0 mL using 0.1 M phosphate buffer. Record the absorbance at 280 and 343 nm.

c. Add 50 µL of DTT solution. Mix well, incubate for 3–5 min at room temperature, and record the absorbance at 343 nm.

d. Using the extinction coefficient at 343 nm of 8.08×10^3/cm/M *(23)*, calculate the amount of pyridine-2-thione liberated during the reduction, which is equivalent to the number of thiols introduced on avidin, using the following equation along with the appropriate extinction coefficient shown in **Table 1**:

$$\frac{\text{Number of thiols}}{\text{avidin}} = \left[\Delta A_{343} / \left(8.08 \infty 10^3\right)\right] \times \left[E_{\text{avidin}}^M / \left(A_{280} - 0.63\Delta A_{343}\right)\right], \quad (1)$$

where ΔA_{343} = change in absorbance at 343 nm, E_{avidin}^M = molar extinction coefficient, and $0.63\Delta A_{343}$ = correction for the absorbance of pyridyldithiopropionate at 280 nm *(23)*.

6. **Equation 1** allows the determination of the average number of moles of enzyme or antibody that can be conjugated with each of avidin (*see* **Note 6**). For a 1:1 protein–avidin conjugate, avidin should be modified with 1.2–1.5 thiols/mol. Thiolated avidin prepared by the above procedure can be stored in the presence of 2 mM sodium azide at 4°C for 4–6 weeks.

3.1.2. Maleimide Derivatization of the Antibody or Enzyme

In this step, which should be completed prior to the deprotection of thiolated avidin, some of the amino groups from the antibody or enzyme are transformed into maleimide groups by reacting with a bifunctional crosslinker, SMCC (*see* **Note 7**).

1. Dissolve or, if already in solution, dialyze the protein in 0.1 M phosphate buffer to obtain a concentration of 2–10 mg/mL. If the protein is an antibody, 11 mg is required to obtain an amount equimolar to 5 mg of avidin (*see* **Note 6**).

2. Prepare a fresh solution of SMCC by dissolving 5 mg in 500 µL of dry DMSO to obtain a 10 mg/mL solution. Vortex or sonicate to ensure that the reagent is completely dissolved.

3. While stirring, add an appropriate amount of SMCC solution to the protein solution to obtain a molar ratio of SMCC to protein of approximately 10. (If 11 mg of an antibody is the protein used, 30 µL of SMCC solution are required.)

Table 1
Molar Extinction Coefficients at 280 nm and Molecular Weights of Avidin and Avidin-Like Proteins

Protein	Molecular weight	E_{Avidin}^M/cm/M
Avidin	66,000	101,640
Deglycosylated avidin/NeutrAvidin	60,000	101,640
CaptAvidin biotin-binding protein	66,000	118,800
Streptavidin	60,000	180,000

4. Continue stirring at room temperature for 1 h.
5. Dialyze the solution in 2 L of 0.1 M phosphate buffer at 4°C for 24 h, with four buffer changes using a membrane with a suitable molecular weight cut-off.

3.1.3. Deprotection of the Avidin Thiol Groups

This procedure is carried out immediately before reacting thiolated avidin with the maleimide derivative of the antibody or enzyme prepared as described in **Subheading 3.1.2.**

1. Dissolve 3 mg of TCEP in 0.3 mL of 0.1 M phosphate buffer.
2. Add 11 µL of TCEP solution to the thiolated avidin solution. Incubate for 15 min at room temperature.

3.1.4. Formation and Purification of the Conjugate

1. Add the thiolated avidin–TCEP mixture dropwise to the dialyzed maleimide-derivatized protein solution with stirring. Continue stirring for 1 h at room temperature, followed by stirring overnight at 4°C.
2. Stop the conjugation reaction by capping residual sulfhydryls with the addition of NEM at a final concentration of 50 µM. Dissolve 6 mg of NEM in 1 mL DMSO and dilute 1:1000 in the conjugate reaction mixture. Incubate for 30 min at room temperature or overnight at 4°C (*see* **Note 8**). The conjugate is now ready for final purification.
3. Concentrate the avidin–protein conjugate mixture to 1–2 mL in a Centricon-30 (Amicon, Beverly, MA, USA) or equivalent centrifuge tube concentrator.
4. Pack appropriate size columns (e.g., 10 × 600 mm for approximately 15 mg of final conjugate) with a degassed matrix suitable for the isolation of the conjugate from unconjugated reagents. If the protein conjugated is an antibody, a matrix such as Bio-Gel A-0.5 m is suitable. For other proteins, Sephadex G-200 or a similar column support may be appropriate, depending on the size of the protein–avidin conjugate.
5. Collect 0.5-mL to 1-mL fractions. The first protein peak to elute contains the conjugate; however, the first or second fraction may contain some aggregates. Analyze each fraction absorbing at 280 nm for biotin binding and assay it for the antibody or enzyme activity. HPLC may also be performed for further purification, if necessary.

3.2. Radioiodination Using IODO-BEADS

The radioiodination procedure (*see* **Note 9**) described here uses IODO-BEADS, which contain the sodium salt of N-chloro-benzenesulfonamide immobilized on nonporous, polystyrene beads. Immobilization of the oxidizing agent allows for easy separation of the latter from the reaction mixture. This method also prevents the use of reducing agents.

1. Wash 6–8 IODO-BEADS twice with 5 mL of PBS. Dry the beads by rolling them on a clean filter paper.
2. Add 500 µL of PBS to the supplier's vial containing 8–10 mCi of carrier-free $Na_{125}I$ or $Na_{131}I$. Place the beads in the same vial and gently mix the contents by swirling. Allow the mixture to sit for 5 min at room temperature with the vial capped.
3. Dissolve or dilute streptavidin in PBS to obtain a final concentration of 1 mg/mL. Add 500 µL of streptavidin solution to the vial containing sodium iodide. Cap the vial immediately and mix the contents thoroughly. Incubate for 20–25 min at room temperature, with occasional swirling (*see* **Note 10**).
4. Carefully remove and save the liquid from the reaction vessel, this is the radioiodinated streptavidin solution. Wash the beads by adding 500 µL of PBS to the reaction vial. Remove the wash solution and add it to the radioiodinated streptavidin.
5. For purification, load the reaction mixture onto a 9- × 200-mm Bio-Gel P-6DG column packed in PBS (0.1% BSA may be added as a carrier to the PBS to reduce loss of streptavidin by adsorption to the column). Elute the column with PBS and collect 0.5-mL fractions. The first set of radioactive fractions (as determined by counting in a γ-ray counter) contains radioiodinated streptavidin, whereas the unreacted radioiodine elutes in the later fractions. Pool the radioiodinated streptavidin fractions.
6. Assessment of protein-associated activity with TCA precipitation:
   a. Dilute a small volume of the pooled radiolabeled streptavidin with saline solution such that 50 µL of the diluted solution has 10^4–10^6 cpm.
   b. Add 50 µL of the diluted streptavidin solution to a 12- × 75-mm glass tube, followed by 500 µL of a 0.1% BSA solution in saline.
   c. For precipitating the proteins, add 500 µL of 10% (w/v) TCA solution in saline.
   d. Incubate the solution for 30 min at room temperature and count the radioactivity of the solution for 10 min ("total counts").
   e. Centrifuge the tube at 500 g for 10 min and carefully discard the supernatant in a radioactive waste container.
   f. Resuspend the pellet in 1 mL of saline and count its radioactivity for 10 min ("bound counts").
   g. The percentage of radioactivity bound to streptavidin is determined using the following equation:

$$[(\text{Bound counts})/(\text{Total counts})] \times 100 = \% \text{ of radioactivity bound to streptavidin} \qquad (2)$$

3.3. Avidin Affinity Matrices

3.3.1. Native Avidin Affinity Matrix

1. Wash 10 mL of sedimented 6% crosslinked agarose with distilled water on a glass or Buchner filter and remove excess water by suction.

2. Dissolve 14 mg of $NaBH_4$ in 7 mL of 1 M NaOH. Add this solution along with 7 mL of 1,4-butanediol-diglycidylether to the washed agarose, with mixing. Allow the reaction to proceed for 10 h or more at room temperature with gentle stirring.

3. Extensively wash the activated gel with distilled water on a supporting filter. The washed gel can be stored in water at 4°C for up to 10 days.

4. Dissolve 50–100 mg of avidin in 10–20 mL of 0.2 M sodium carbonate, pH 9.5, and suspend the sedimented activated agarose gel in the same buffer to obtain a workable slurry.

5. Slowly drip the agarose slurry into the stirred protein solution and allow the binding to take place at room temperature for 2 days with continuous gentle mixing.

6. Wash the avidin–agarose mixture in PBS until the filtrate shows no absorbance at 280 nm. Store at 4°C in the presence of 0.02% sodium azide.

3.3.2. Monomeric Avidin Affinity Matrix

1. Filter the avidin–agarose matrix (*see* **Subheading 3.3.1., step 6**) on a glass or Buchner filter (or pack in a column) and wash four times with 2 vol of 6 M guanidine HCl in 0.2 M KCl, pH 1.5, to dissociate the tetrameric avidin.

2. Thoroughly wash the gel with 0.2 M potassium phosphate, pH 7.5, and suspend in 10 mL of the same buffer.

3. Add 3 mg of solid succinic anhydride to succinylate the monomeric avidin and incubate for 1 h at room temperature with gentle stirring.

4. Wash the gel with 0.2 M potassium phosphate, pH 7.5, pack in a column, and saturate the binding sites by running through three volumes of 1 mM biotin dissolved in the same buffer.

5. Remove biotin from the low-affinity binding sites by washing the column with 0.2 M glycine HCl, pH 2.0.

6. Store the column equilibrated in 20 mM sodium phosphate, 0.5 M NaCl, 0.02% sodium azide, pH 7.5. The column is now ready to use.

7. Load the column with the mixture to be purified. Elute any unbound protein by adding 20 mM sodium phosphate, 0.5 M NaCl, pH 7.5. Add biotin to the same buffer to obtain a final concentration of 0.8 mM to elute the biotinylated compound.

8. Regenerate the column after each run by washing with 0.2 M glycine HCl, pH 2.0.

4. Notes

1. A detailed procedure for the conjugation of fluorophores to antibodies has been recently published *(24)*. This protocol can be modified for conjugation of fluorophores to avidin or avidin-related proteins by using a dye to avidin molar ratio of 5–8:1.

2. The conjugation reaction for oligonucleotides synthesized with a disulfide containing a protecting group should be performed under nitrogen or argon. Deprotect the disulfide of the oligonucleotide using DTT. Add 1 mg of DTT to 140 μL of a 6 μM oligonucleotide (21–33 mer) solution in 0.1 M phosphate buffer containing 5 mM ethylenediaminetetraacetic acid. Stir the solution at 37°C for 0.5 h. Purify the reaction mixture using a disposable desalting column. Combine the oligonucleotide-containing fractions with thiolated avidin prepared as described in **Subheading 3.1.1.** It should be noted that, in this case, conjugation occurs through the formation of a disulfide bond instead of a thioether bond. Disulfides are sensitive to reducing agents; however, they make reasonably stable conjugates, useful in most applications *(25)*. Purify the conjugate as outlined in **Subheading 3.1.4.**

3. Using a molar ratio of SPDP to avidin of 5 yields 1–2 protected sulfhydryls per molecule of avidin. This range of thiols per mole is found to produce the best yield of a 1:1 conjugate.

4. Buffer and pH: The entire procedure for preparation of conjugates through thioether bonds can be performed at pH 7.5. (Note: Organic buffers containing amines, such as Tris, are unsuitable.) Antibodies or enzymes in PBS can be prepared for reaction with SMCC by adding 1/10 vol of 1 M sodium bicarbonate solution. This step eliminates dialysis and consequent dilution of the protein. Presence of azide at concentrations above 0.1% may interfere with the reaction of the protein with SMCC or of avidin with SPDP. IgM antibodies denature above pH 7.2. They can, however, be conjugated in PBS at pH 7.0 by increasing the molar ratio of maleimide to antibody.

5. Peptides (20–25 amino acids) containing a single cysteine can also be conjugated to thiolated avidin by modifying the procedure described in **Subheading 3.1.** and performing the reaction under argon or nitrogen *(26)*. Peptide–avidin conjugate formation described here also involves the formation of a disulfide bond. For conjugation with 5 mg avidin, dissolve 1.6 mg of a lyophilized cysteine-containing peptide in 900 μL water/methanol (2:1 v/v) using 50 mM NaOH (a few microliters at a time) to improve solubility. Immediately prior to use, cleave any cystine-bridged homodimer that may be present by the addition of TCEP solution (10 mg/mL in 0.1 M phosphate buffer) to obtain a TCEP to peptide ratio of 3. Incubate for 15 min at room temperature. Purify the peptide–TCEP mixture using a disposable desalting column. Combine the peptide-containing fractions with thiolated avidin prepared as described in **Subheading 3.1.1.** Purify the conjugate as described in **Subheading 3.1.4.**

6. Avidin and antibody or enzyme concentration: The concentration of avidin as well as that of the protein to be conjugated should be 2–10 mg/mL. The crosslinking efficiency and, consequently, the yield of the conjugate decreases at lower concentrations of the thiolated avidin and maleimide-derivatized protein. To obtain 1:1 conjugates, equimolar concentrations of avidin and the protein are desirable. However, most methods of conjugation will generate conjugates of different sizes, following the Poisson distribution. The size range obtained with the method

described here is much narrower because the number of proteins reacting with each mole of avidin can be regulated by the degree of thiolation of avidin.

7. It is essential that the procedure described in **Subheading 3.1.2.** be performed approximately 24 h before the procedure described in **Subheading 3.1.3.**, because the deprotected thiolated avidin and the maleimide derivative of the protein are unstable. Purification of the maleimide-derivatized protein by size exclusion chromatography can be performed more rapidly than dialysis; however, the former leads to dilution of the protein and a decrease in the yield of the conjugate.

8. If the molecule being conjugated to avidin is β-galactosidase or other free thiol-containing oligonucleotide or protein, NEM treatment is not performed.

9. Radioiodination of streptavidin uses procedures similar to those used for stable nuclides. However, some distinct differences remain, as radioiodinations are performed in dilute solutions. Also, the radioiodination mixture contains minor impurities formed during the preparation and purification of the radionuclide. Thus, optimization of reaction parameters is essential for performing radioiodi-nation. This reaction is carried out in small volumes; it is therefore essential to ensure adequate mixing at the outset of the reaction. Inadequate mixing is often responsible for poor radioiodination yield.

10. Specific activity using the method described in **Subheading 3.2.** is usually in the range of 10–50 mCi/mg, and the protein-bound radioactivity obtained is >95%. Higher specific activity can be achieved by increasing the reaction time (*see* **Subheading 3.2., step 3**) in by using more beads, or by increasing the amount of radioiodine. However, one must bear in mind that at longer incubation times, the risk of damage to streptavidin is greater.

11. Storage and stability of avidin conjugates: Most avidin conjugates can be stored at 4 or –20°C after lyophilization. Because of the variation in antibody structure, there is no general rule on the best method to store avidin–antibody conju-gates, and the best conditions are determined experimentally. Aliquoting in small amounts and freezing is generally satisfactory. Radiolabeled streptavidin is aliquoted (~100 mL/tube) and stored at 4 or –20°C until use.

References

1. Green, N. M. (1975) Avidin, in *Advances in Protein Chemistry*, vol. 29 (Anfinsen, C. M., Edsall, J. T., and Richards, F. M., eds.), Academic Press, New York, pp. 85–133.
2. Bayer, E. A. and Wilchek, M. (1980) The use of the avidin–biotin complex as a tool in molecular biology. *Methods Biochem. Anal.* **26,** 1–45.
3. Niemeyer, C. M., Sano, T., Smith, C. L., and Cantor, C. R. (1994) Oligonucleotide-directed self-assembly of proteins: semisynthetic DNA–streptavidin hybrid molecules as connectors for the generation of macroscopic arrays and the construction of supramolecular bioconjugates. *Nucleic Acids Res.* **22,** 5330–5339.
4. Paganelli, G., Belloni, C., Magnani, P., Zito, F., Pasini, A., Sassi, I., Meroni, M., Mariani, M., Vignali, M., Siccardi, A. G., and Fazio, F. (1992) Two-step tumor

targetting in ovarian cancer patients using biotinylated monoclonal antibodies and radioactive streptavidin. *Eur. J. Nucl. Med.* **19,** 322–329.

5. van Osdol, W. W., Sung, C., Dedrick, R. L., and Weinstein, J. N. (1993) A distributed pharmacokinetic model of two-step imaging and treatment protocols: application to streptavidin-conjugated monoclonal antibodies and radiolabeled biotin. *J. Nucl. Med.* **34,** 1552–1564.

6. Markwell, M. A. K. (1982) A new solid-state reagent to iodinate proteins. I. Conditions for the efficient labeling of antiserum. *Anal. Biochem.* **125,** 427–432.

7. Salacinski, P. R. P., McLean, C., Sykes, J. E. C., Clement-Jones, V. V., and Lowry, P. J. (1981) Iodination of proteins, glycoproteins, and peptides using a solid-phase oxidizing agent, 1,3,4,6,-tetrachloro-3α,6α-diphenyl glycoluril (iodogen). *Anal. Biochem.* **117,** 136–146.

8. Mock, D. M. (1990) Sequential solid-phase assay for biotin based on 125I-labeled avidin. *Methods Enzymol.* **184,** 224–233.

9. Bolton, A. E. and Hunter, W. M. (1973) The labeling of proteins to high specific radioactivity by conjugation to an ^{125}I-containing acylating agent. Applications to the radioimmunoassay. *Biochem. J.* **133,** 529–539.

10. Vaidyanathan, G., Affleck, D. J., and Zalutsky, M. R. (1993) Radioiodination of proteins using N-succinimidyl 4-hydroxy-3-iodobenzoate. *Bioconjugate Chem.* **4,** 78–84.

11. Vaidyanathan, G. and Zalutsky, M. R. (1990) Radioiodination of antibodies via N-succinimidyl 2,4-dimethoxy-3-(trialkylstannyl)benzoates. *Bioconjugate Chem.* **1,** 387–393.

12. Hylarides, M. D., Wilbur, D. S., Reed, M. W., Hadley, S. W., Schroeder, J. R., and Grant, L. M. (1991) Preparation and *in vivo* evaluation of an N-(p-[125I]iodophenethyl)maleimide-antibody conjugate. *Bioconjugate Chem.* **2,** 435–440.

13. Arano, Y., Wakisaka, K., Ohmomo, Y., Uezono, T., Mukai, T., Motonari, H., Shiono, H., Sakahara, H., Konishi, J., Tanaka, C., and Yokoyama, A. (1994) Maleimidoethyl 3-(tri-*n*-butylstannyl)hippurate: a useful radioiodination reagent for protein radiopharmaceuticals to enhance target selective radioactivity local-ization. *J. Med. Chem.* **37,** 2609–2618.

14. Willner, D., Trail, P. A., Hofstead, S. J., Dalton King, H., Lasch, S. J., Braslawsky, G. R., Greenfield, R. S., Kaneko, T., and Firestone, R. A. (1993) (6-Maleimidocaproyl)hydrazone of doxorubicin—A new derivative for the prepa-ration of immunoconjugates of doxorubicin. *Bioconjugate Chem.* **4,** 521–527.

15. Arnold, L. J. Jr. (1985) Polylysine-drug conjugates. *Methods Enzymol.* **112,** 270–285.

16. Pietersz, G. A. and McKenzie, I. F. (1992) Antibody conjugates for the treatment of cancer. *Immunol. Rev.* **129,** 57–80.

17. Orr, G. A. (1981) The use of the 2-iminobiotin–avidin interaction for the selective retrieval of labeled plasma membrane components. *J. Biol. Chem.* **256,** 761–766.

18. Dimroth, P. (1986) Preparation, characterization, and reconstitution of oxaloacetate decarboxylase from *Klebsiella aerogenes*, a sodium pump. *Methods Enzymol.* **125,** 530–540.

19. Dean, P. D. G., Johnson, W. S., and Middle, F. S. (1985) Activation procedures, in *Affinity Chromatography. A Practical Approach*, IRL, Washington, DC, pp. 34,35.

20. Kohanski, R. A. and Lane, D. (1990) Monovalent avidin affinity columns. *Methods Enzymol.* **184,** 194–220.

21. Chilkoti, A., Schwartz, B. L., Smith, R. D., Long, C. J., and Stayton, P. S. (1995) Engineered chimeric streptavidin tetramers as novel tools for bioseparations and drug delivery. *Biotechnology (NY)* **13,** 1198–1204.

22. Wong, S. S. (1991) Reactive groups of proteins and their modifying agents, in *Chemistry of Protein Conjugation and Crosslinking*, CRC, Boston, MA, pp. 7–48.

23. Carlsson, J., Drevin, H., and Axen, R. (1978) Protein thiolation and reversible protein–protein conjugation. N-Succinimidyl 3-(2-pyridyldithio)propionate, a new heterobifunctional reagent. *Biochem. J.* **173,** 723–737.

24. Haugland, R. P. (1995) Coupling of monoclonal antibodies with fluorophores, in *Methods in Molecular Biology*, vol. 45 (Davis, W. C., ed.), Humana, Totowa, NJ, pp. 205–221.

25. Kronick, M. N. and Grossman, P. D. (1983) Immunoassay techniques with fluorescent phycobiliprotein conjugates. *Clin. Chem.* **29,** 1582–1586.

26. Bongartz, J.-P., Aubertin, A.-M., Milhaud, P. G., and Lebleu, B. (1994) Improved biological activity of antisense oligonucleotides conjugated to a fusogenic peptide. *Nucleic Acids Res.* **22,** 4681–4688.

2

Coupling of Antibodies With Biotin

Rosaria P. Haugland and Wendy W. You

1. Introduction

The avidin–biotin bond is the strongest known biological interaction between a ligand and a protein (K_d = 1.3 × 10^{-15} M at pH 5.0) *(1)*. The affinity is so high that the avidin–biotin complex is extremely resistant to any type of denaturing agent *(2)*. Biotin (*see* **Fig. 1**) is a small, hydrophobic molecule that functions as a coenzyme of carboxylases *(3)*. It is present in all living cells. Avidin is a tetrameric glycoprotein of 66,000–68,000 molecular weight, found in egg albumin and in avian tissues. The interaction between avidin and biotin occurs rapidly, and the stability of the complex has prompted its use for in situ attachment of labels in a broad variety of applications, including immunoassays, DNA hybridization *(4–6)*, and localization of antigens in cells and tissues *(7)*. Avidin has an isoelectric point of 10.5. Because of its positively charged residues and its oligosaccharide component, consisting mostly of mannose and glucosamine *(8)*, avidin can interact nonspecifically with negative charges on cell surfaces and nucleic acids, or with membrane sugar receptors. At times, this causes background problems in histochemical and cytochemical applications. Streptavidin, a near-neutral, biotin-binding protein *(9)* isolated from the culture medium of *Streptomyces avidinii*, is a tetrameric nonglycosylated analog of avidin with a molecular weight of about 60,000. Like avidin, each molecule of streptavidin binds four molecules of biotin, with a similar dissociation constant. The two proteins have about 33% sequence homology, and tryptophan residues seem to be involved in their biotin-binding sites *(10,11)*. In general, streptavidin gives less background problems than avidin. This protein, however, contains a tripeptide sequence Arg-Tyr-Asp (RYD) that apparently mimics the binding sequence of fibronectin Arg-Gly-Asp (RGD), a universal recognition

From: *Methods in Molecular Biology, vol. 418: Avidin-Biotin Interactions, Methods and Applications*
Edited by: R. J. McMahon © Humana Press, Totowa, NJ

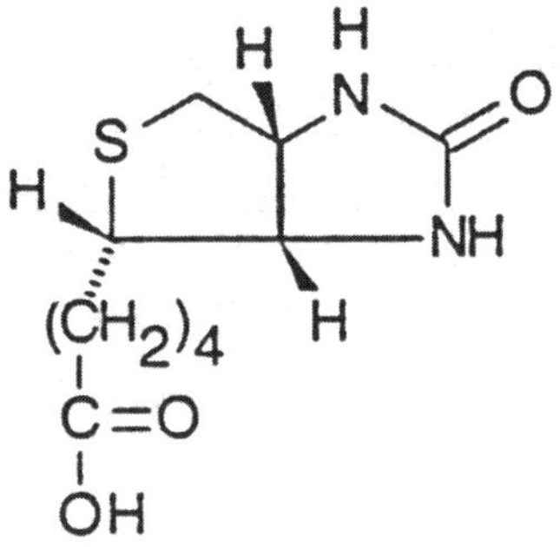

H H
S N O
H
—NH
(CH₂)₄ H
C=O
OH

Biotin MW 244.31

Fig. 1. Structure of biotin.

domain of the extracellular matrix that specifically promotes cell adhesion. Consequently, the streptavidin–cell-surface interaction causes high background in certain applications *(12)*.

As an alternative to both avidin and streptavidin, a chemically modified avidin, NeutraLite™ avidin (NeutraLite is a trademark of Belovo Chemicals, Bastogne, Belgium), has recently become available. NeutraLite™ avidin consists of chemically deglycosylated avidin, which has been modified to reduce the isoelectric point to a neutral value, without loss of its biotin-binding properties and without significant change in the lysines available for derivatization *(13)*. (Fluorescent derivatives and enzyme conjugates of NeutraLite™ avidin, as well as the unlabeled protein, are available from Molecular Probes, Eugene, OR, USA.)

As shown in **Fig. 1**, biotin is a relatively small and hydrophobic molecule. The addition to the carboxyl group of biotin of one (X) or two (XX) aminohexanoic acid "spacers" greatly enhances the efficiency of formation of the complex between the biotinylated antibody (or other biotinylated protein) and the avidin–probe conjugate, where the probe can be a fluorochrome or an enzyme *(14,15)*. Each of these 7- or 14-atom spacer arms has been shown to improve the ability of biotin derivatives to interact with the binding cleft of avidin. The comparison between streptavidin binding activity of proteins biotinylated with biotin-X or with biotin-XX (labeled with same number of moles of biotin/mol of protein) has been performed in our laboratory (*see* **Fig. 2**). No difference was found between the avidin and streptavidin–horseradish peroxidase conjugates in their ability to bind biotin-X or biotin-XX. However, biotin-XX gave consistently higher titers in enzyme-linked immunosorbent assays, using biotinylated goat antimouse Ig (GAM), bovine serum albumin (BSA), or protein A (results with avidin and with protein A are not presented here). Even nonroutine conju-

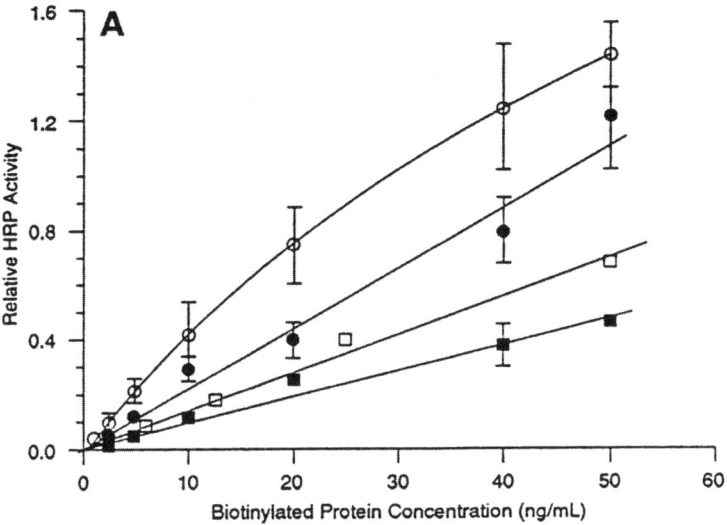

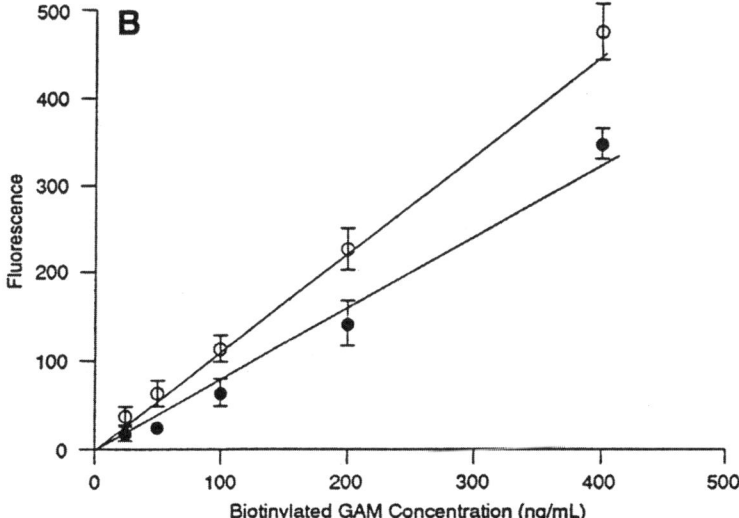

Fig. 2. (A) ELISA-type assay comparing the binding capacity of BSA and GAM biotinylated with biotin-*X* or biotin-*XX*. The assay was developed using streptavidin-HRP conjugate (0.2 μg/mL) and *o*-phenylenediamine dihydro-chloride (OPD). The number of biotin/mol was: 4.0 biotin-*X*/GAM (●), 4.4 biotin-*XX*/GAM (○), 6.7 biotin-*X*/BSA (■), and 6.2 biotin-*XX*/BSA (□). Error bars on some data points have been omitted for clarity. (B) Similar assay using GAM biotinylated with biotin-*X* (●) or biotin-*XX* (○). The assay was developed with streptavidin–R-phycoerythin conjugate (25 μg/mL using a Millipore CytoFluor™ fluorescence microtiter plate reader).

gations performed in our laboratory have consistently yielded excellent results using biotin-XX.

Biotin, biotin-X, and biotin-XX have all been derivatized for conjugation to amines or thiols of proteins and aldehyde groups of glycoproteins or other polymers. The simplest and most popular biotinylation method is to label the ε-amino groups of lysine residues with a succinimidyl ester of biotin. Easy-to-use biotinylation kits that facilitate the biotinylation of 1–2 mg of protein or oligonucleotides are commercially available *(16)*. One kit for biotinylating smaller amounts of protein (0.1–3 mg) utilizes biotin-XX sulfosuccinimidyl ester *(17)*. This compound is water-soluble and allows for the efficient labeling of dilute protein samples. Another kit uses biotin-X 2,4-dinitrophenyl-X-lysine succinimidyl ester (DNP-biocytin) as the biotinylating reagent. DNP-biocytin was developed by Molecular Devices (Menlo Park, CA, USA) for their patented Threshold-Immunoligand System *(18)*. DNP-biocytin permits the direct measurement of the degree of biotinylation of the reaction product by using the molar extinction coefficient of DNP (15,000/cm at 364 nm). Conjugates of DNP-biocytin can be probed separately or simultaneously using either anti-DNP antibodies or avidin/streptavidin; this flexibility is useful when combining techniques, such as fluorescence and electron microscopy. Biotin iodoacetamide or maleimide, which could biotinylate the reduced sulfhydryls located at the hinge region of antibodies, is not usually used for this purpose. More examples in the literature describe biotinylation of antibodies with biotin hydrazide at the carbohydrate prosthetic group, located in the Fc portion of the molecule, relatively removed from the binding site. Conjugation of carbohydrates with hydrazides requires the oxidation of two adjacent hydroxyls to aldehydes and optional stabilization of the reaction with cyanoborohydride *(19)*.

Because of its strength, the interaction between avidin and biotin cannot be used for preparing matrices for affinity column purification, unless columns prepared with avidin monomers are used *(20)*. The biotin analog, iminobiotin, which has a lower affinity for avidin, can be used for this purpose *(21,22)*. Iminobiotin in reactive form is commercially available, and the procedure for its conjugation is identical to that used for biotin. Detailed, practical protocols for biotinylating antibodies at the lysine or at the carbohydrate site and a method to determine the degree of biotinylation are described in detail in this chapter (*see* **Notes 1–10** for review of factors that affect optimal conjugation and yield of biotinylated antibodies).

2. Materials

2.1. Conjugation with Amine-Reactive Biotin

1. Reaction buffer: 1 M sodium bicarbonate, stable for about 2 weeks when refrigerated. Dissolve 8.3 g of $NaHCO_3$ in 100 mL of distilled water. The pH will be

about 8.3. Dilute 1:10 before using to obtain a 0.1 M solution. Alternate reaction buffer: 0.1 M sodium phosphate, pH 7.8. Dissolve 12.7 g Na_2HPO_4 and 1.43 g NaH_2PO_4 in 800 mL of distilled water. Adjust pH to 7.8 if necessary. Bring the volume to 1000 mL. This buffer is stable for 2 months when refrigerated.

2. Anhydrous dimethylformamide (DMF) or dimethyl sulfoxide (DMSO).
3. Phosphate-buffered saline (PBS): Dissolve 1.19 g of K_2HPO_4, 0.43 g of KH_2PO_4· H_2O, and 8.8 g NaCl in 800 mL of distilled water, adjust the pH to 7.2 if necessary or to the desired pH, and bring the volume to 1000 mL with distilled water.
4. Disposable desalting columns or a gel filtration column: Amicon GH-25 and Sephadex G-25 or the equivalent, equilibrated with PBS or buffer of choice.
5. Good quality dialysis tubing as an alternative to the gel filtration column when derivatizing small quantities of antibody.
6. Biotin, biotin-X, or biotin-XX succinimidyl ester: As with all succinimidyl esters, these compounds should be stored well desiccated in the freezer.

2.2. Conjugation with Biotin Hydrazide at the Carbohydrate Site

1. Reaction buffer: 0.1 M acetate buffer, pH 6.0. Dilute 5.8 mL acetic acid in 800 mL distilled water. Bring the pH to 6.0 with 5 M NaOH and the volume to 1000 mL. The buffer is stable for several months when refrigerated.
2. Sodium metaperiodate, 20 mM: Dissolve 43 mg of $NaIO_4$ in 10 mL of reaction buffer, protecting from light. Use fresh.
3. Biotin-X hydrazide or biotin-XX hydrazide.
4. DMSO.
5. Optional: 100 mM sodium cyanoborohydride, freshly prepared. Dissolve 6.3 mg of $NaBH_3CN$ in 10 mL of 0.1 mM NaOH.

2.3. Determination of the Degree of Biotinylation

1. 4′ Hydroxyazobenzene-2-carboxylic acid (HABA) (10 mM) in 10 mM NaOH.
2. 50 mM Sodium phosphate and 150 mM NaCl, pH 6.0. Dissolve 0.85 g of Na_2HPO_4 and 6.07 g of NaH_2PO_4 in 800 mL of distilled water. Add 88 g of NaCl. Bring the pH to 6.0 if necessary and the volume to 1000 mL.
3. Avidin, 0.5 mg/mL, in 50 mM sodium phosphate and 150 mM NaCl, pH 6.0.
4. Biotin, 0.25 mM, in 50 mM sodium phosphate, and 150 mM NaCl, pH 6.0.

3. Methods
3.1. Conjugation with Amine-Reactive Biotin

1. Dissolve the antibody, if lyophilized, at approximately 5–15 mg/mL in either of the two reaction buffers described in **Subheading 2.1.** If the antibody to be conjugated is already in solution in 10–20 mM PBS, without azide, the pH necessary for the reaction can be obtained by adding 1/10 vol of 1 M sodium bicarbonate. IgM should be conjugated in PBS, pH 7.2 (*see* **Note 3**).

2. Calculate the amount of a 10 mg/mL biotin succinimidyl ester solution (biotin-SE) needed to conjugate the desired quantity of antibody at the chosen biotin/antibody molar ratio, according to the following formula:

$$\text{(mL of 10 mg/mL biotin-SE)} = \{[(\text{mg antibody} \times 0.1)/\text{mol wt of antibody}] \times R \times \text{mol wt of biotin-SE})\}, \qquad (1)$$

where R = molar incubation ratio of biotin/protein. For example, using 5 mg of IgG and a 10:1 molar incubation ratio of biotin-XX-SE, **Eq. 1** yields:

$$\text{(mL of 10 mg/mL biotin-XX-SE)} = \{ [(5 \times 0.1)/145{,}000] \times (10 \times 568)\}$$
$$= 0.02 \text{ mL} \qquad (2)$$

3. Weigh 3 mg or more of the biotin-SE of choice and dissolve it in 0.3 mL or more of DMF or DMSO to obtain a 10 mg/mL solution. It is essential that this solution be prepared immediately before starting the reaction, as the succinimidyl esters or any amine-reactive reagents hydrolyze quickly in solution. Any remaining solution should be discarded.

4. While stirring, slowly add the amount of 10 mg/mL solution, calculated according to the formula given in **step 2**, to the antibody prepared as described in **step 1**, mixing thoroughly.

5. Incubate this reaction mixture at room temperature for 1 h with gentle stirring or shaking.

6. The antibody conjugate can be purified on a gel filtration column or by dialysis. When working with a few milligrams of dilute antibody solution, care should be taken not to dilute the antibody further. In this case, dialysis is a very simple and effective method to eliminate unreacted biotin. A few milliliters of antibody solution can be effectively dialyzed in the cold against 1 L of buffer with three to four changes. Small amounts of concentrated antibody can be purified on a prepackaged desalting column equilibrated with the preferred buffer, following the manufacturer's directions. Five or more milligrams of antibody can be purified on a gel filtration column. The dimensions of the column will have to be proportional to the volume and concentration of the antibody. For example, for 5–10 mg of antibody in 1 mL solution, a column with a bed volume of 10 × 300 mm will be adequate. To avoid denaturation, dilute solutions of biotinylated antibodies should be stabilized by adding BSA at a final concentration of 0.1–1%.

3.2. Conjugation with Biotin Hydrazide at the Carbohydrate Site

1. It is essential that the following entire procedure be carried out with the sample completely protected from light (*see* **Note 9**).

2. Dissolve antibody (if lyophilized) or dialyze solution of antibody to obtain a 2–10 mg/mL solution in the reaction buffer described in **Subheading 2.1., item 1**. Keep at 4°C.

3. Add an equal volume of cold metaperiodate solution. Incubate the reaction mixture at 4°C for 2 h in the dark.
4. Dialyze overnight against the same buffer protecting from light, or, if the antibody is concentrated, desalt on a column equilibrated with the same buffer. This step removes the iodate and formaldehyde produced during oxidation.
5. Dissolve 10 mg of the biotin hydrazide of choice in 0.25 mL of DMSO to obtain a 40 mg/mL solution, warming if needed. This will yield a 107 mM solution of biotin-X hydrazide or an 80 mM solution of biotin-XX hydrazide. These solutions are stable for a few weeks.
6. Calculate the amount of biotin hydrazide solution needed to obtain a final concentration of approximately 5 mM and add it to the oxidized antibody. When using biotin-X hydrazide, 1 vol of hydrazide should be added to 20 vol of antibody solution. When using biotin-XX hydrazide, 1 vol of hydrazide should be added to 15 vol of antibody solution.
7. Incubate for 2 h at room temperature with gentle stirring.
8. This step is optional. The biotin hydrazone–antibody conjugate formed in this reaction (**steps 6** and **7**) is considered by some researchers to be relatively unstable. To reduce the conjugate to a more stable, substituted hydrazide, treat the conjugate with sodium cyanoborohydride at a final concentration of 5 mM by adding a 1/20 vol of a 100 mM stock solution. Incubate for 2 h at 4°C (*see* **Note 5**).
9. Purify the conjugate by any of the methods described for biotinylating antibodies at the amine site (*see* **Subheading 3.1., step 6**).

3.3. Determination of the Degree of Biotinylation

The dye HABA interacts with avidin yielding a complex with an absorption maximum at 500 nm. Biotin, because of its higher affinity, displaces HABA, causing a decrease in absorbance at 500 nm proportional to the amount of biotin present in the assay.

1. To prepare a standard curve, add 0.25 mL of HABA reagent to 10 mL of avidin solution. Incubate for 10 min at room temperature and record the absorbance at 500 nm of 1 mL avidin–HABA complex with 0.1 mL buffer, pH 6.0. Distribute 1 mL of the avidin–HABA complex into six test tubes. Add to each the biotin solution in a range of 0.005–0.10 mL. Bring the final volume to 1.10 mL with pH 6.0 buffer and record the absorbance at 500 nm of each concentration point. Plot a standard curve with the nanomoles of biotin versus the decrease in absorbance at 500 nm. An example of a standard curve is illustrated in **Fig. 3**.
2. To measure the degree of biotinylation of the sample, add an aliquot of biotinylated antibody of known concentration to 1 mL of avidin–HABA complex. For example, add 0.05–0.1 mL of biotinylated antibody at 1 mg/mL to 1 mL of avidin–HABA mixture. Bring the volume to 1.10 mL, if necessary, incubate for 10 min, and measure the decrease in absorbance at 500 nm.
3. Deduct from the standard curve the nanomoles of biotin corresponding to the observed change in absorbance. The ratio between nanomoles of biotin and

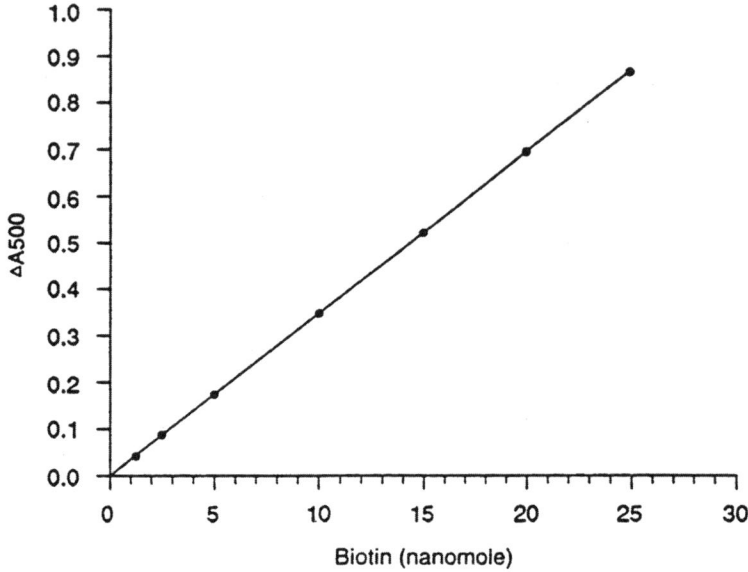

Fig. 3. Examples of standard curve for biotin assay with avidin-HABA reagent, obtained as described in **Subheading 3.3**.

nanomoles of antibody used to displace HABA represents the degree of biotinylation, as seen from the following equation:

$$[(\text{nmol biotin} \times 145{,}000 \times 10^{-6})/$$
$$(\text{mg/mL antibody} \times 0.1 \text{ mL})] = (\text{mol of biotin/mol of antibody}), \tag{3}$$

where 145,000 represents the molecular weight of the antibody and 0.1 mL is the volume of 1 mg/mL of biotinylated antibody sample.

4. Notes

4.1. Factors that Influence the Biotinylation Reaction

1. Protein concentration: As in any chemical reaction, the concentration of the reagents is a major factor in determining the rate and the efficiency of the coupling. Antibodies at a concentration of 5–20 mg/mL will give better results; however, it is often difficult to have such concentrations or even such quantities available for conjugation. Nevertheless, the antibody should be as concentrated as possible. In the case of solutions of antibody <2–3 mg/mL, the molar ratio of biotinylating reagent (or of both the oxidizing and biotinylating reagent, in the case of labeling the carbohydrate region) should be increased. It is also essential that the antibody solutions do not contain gelatin or BSA, which are often added

to stabilize dilute solutions of antibodies. These proteins, generally present at a 1% concentration, will also react with biotinylating reagents.

2. pH: The reactivity of amines increases at basic pH. Unfortunately, so does the rate of hydrolysis of succinimidyl esters. We have found that the best pH for biotinylation of the ε-amino groups of lysines is 7.5–8.3. IgM antibodies, which denature at basic pH, can be biotinylated at pH 7.2 by increasing the molar ratio of the biotinylating reagent to antibody to at least 20. The optimum pH for oxidation and conjugation with hydrazides is 5.5–6.0.

3. Buffer: Bicarbonate or phosphate buffers are suitable for biotinylation. Organic buffers, such as Tris, which contain amines, should be avoided, because they react with amino-labeling reagents or interfere with the reaction between aldehydes and hydrazides. However, HEPES and EPPS, which contain tertiary amines, are suitable. Antibodies dissolved in 10–20 mM PBS can be readily prepared for conjugation at the lysine site by adding 1/10–1/5 of the volume of 1 M sodium bicarbonate. As noted, because IgM antibodies are unstable in basic solution, biotinylation at the ε-amino group of lysines should be attempted in PBS or equivalent buffer at pH 7.2. Reactions of antibodies with periodate and biotin hydrazide can be performed in PBS at pH 7.0 or in acetate buffer, pH 6.0 (*see* **Subheading 2.2.**).

4. Temperature: Biotinylations at the amino group sites are run at room temperature, at the carbohydrate site at 0–4°C.

5. Time: Succinimidyl ester derivatives will react with a protein within 1 h. Periodate oxidation will require 2 h at pH 6.0. Reaction with biotin hydrazide can be performed in a few hours. Stabilization with cyanoboro-hydride requires <2 h.

6. Desired degree of biotinylation and stability of the conjugate: Reaction of an antibody with biotin does not significantly alter the size or charge of the molecule. However, because of the size of avidin or its analogs (molecular weight = 60,000–68,000), an increase in the number of biotins per antibody will not necessarily increase the number of avidins capable of reacting with one antibody molecule. Because biotin, biotin-X, and biotin-XX are hydrophobic molecules, a high degree of biotinylation might increase the background or might destabilize the antibody. To obtain a degree of biotinylation of about 3–7 biotins/IgG, generally a molar ratio of 15 mol of amino biotinylating reagent/mol of protein is used. When the concentration of the antibody is <3 mg/mL, this ratio should be increased. The amount of increase should be determined experimentally, because the reactivity of the lysines available for conjugation varies for each antibody. This could become a significant factor, especially at low antibody concentrations.

The succinimidyl esters or hydrazides of biotin, biotin-X, and biotin-XX exhibit similar degrees of reactivity, and the choice is up to the researcher. In general, the longer spacer arm in biotin-XX should be advantageous (*see* **Fig. 2**). The overall stability of biotinylated monoclonal antibodies derivatized with a moderate number of biotin should be similar to the stability of the native antibody, and the storage conditions also should be the same.

4.2. Factors that Affect Antibodies

7. Most antibodies can withstand biotinylation with minimal change in activity and stability, especially if the degree of biotinylation is about 3–6 biotins/mol.

8. Biotin or any of its longer chain derivatives do not contribute to the absorbance of the antibody at 280 nm. Consequently, the concentration of the antibody can be measured by using $A^{1\%}_{1cm}=0$ 14 at 280 nm.

9. It is essential that the entire procedure for biotinylation of antibodies at the carbohydrate site (*see* **Subheading 3.2.**) be performed in the dark, protected from light.

10. It should be noted that dry milk, serum, and other biological fluids contain biotin and, consequently, they should not be used as blocking agents in systems where blocking is required.

References

1. Green, N. M. (1963) Avidin. 3. The nature of the biotin binding site. *Biochem. J.* **89,** 599–609.

2. Green, N. M. (1963) Avidin. 4. Stability at extremes of pH and dissociation into subunits by guanidine hydrochloride. *Biochem. J.* **89,** 609–620.

3. Knappe, J. (1970) Mechanism of biotin action. *Annu. Rev. Biochem.* **39,** 757–776.

4. Wilchek, M. and Bayer, E. A. (1988) The avidin–biotin complex in bioanalytical applications. *Anal. Biochem.* **171,** 1–32.

5. Wilchek, M. and Bayer, E. A. (1990) Avidin–biotin technology, in *Methods in Enzymology*, vol. 184, Academic Press, New York, pp. 213–217.

6. Levi, M., Sparvoli, E., Sgorbati, S., and Chiantante, D. (1990) Biotin–streptavidin immunofluorescent detection of DNA replication in root meristems through Brd Urd incorporation: cytological and microfluorimetric applications. *Physiol. Plant.* **79,** 231–235.

7. Armstrong, R., Friedrich, V. L., Jr., Holmes, K. V., and Dubois-Dalcq, M. (1990) *In vitro* analysis of the oligodendrocyte lineage in mice during demyelination and remyelination. *J. Cell Biol.* **111,** 1183–1195.

8. Bruch, R. C. and White, H. B., III (1982) Compositional and structural heterogeneity of avidin glycopeptides. *Biochemistry* **21,** 5334–5341.

9. Hiller, Y., Gershoni, J. M., Bayer, E. A., and Wilchek, M. (1987) Biotin binding to avidin: oligosaccharide side chain not required for ligand association. *Biochem. J.* **248,** 167–171.

10. Green, N. M. (1975) Avidin, in *Advances in Protein Chemistry*, vol. 29 (Anfinsen, C. B., Edsall, J. T., and Richards, F. M., eds.), Academic Press, New York, pp. 85–133.

11. Chaiet, L. and Wolf, F. J. (1964) The properties of streptavidin, a biotin-binding protein produced by *Streptomycetes*. *Arch. Biochem. Biophys.* **106,** 1–5.

12. Alon, R., Bayer, E. A., and Wilcheck, M. (1990) Streptavidin contains an Ryd sequence which mimics the RGD receptor domain of fibronectin. *Biochem. Biophys. Res. Commun.* **170,** 1236–1241.

13. Wilchek, M. and Bayer, E. A. (1993) Avidin–biotin immobilization systems, in *Immobilized Macromolecules: Application Potentials* (Sleytr, U. B., ed.), Springer-Verlag, New York, pp. 51–60.

14. Gretch, D. R., Suter, M., and Stinski, M. F. (1987) The use of biotinylated monoclonal antibodies and streptavidin affinity chromatography to isolate Herpes virus hydrophobic proteins or glycoproteins. *Anal. Biochem.* **163,** 270–277.

15. Hnatowich, D. J., Virzi, F., and Rusckowski, M. (1987) Investigations of avidin and biotin for imaging applications. *J. Nucl. Med.* **28,** 1294–1302.

16. Haugland, R. P. (1996) Biotin and haptens, in *Handbook of Fluorescent Probes and Research Chemicals,* 6th ed. (Spence, M., ed.), Molecular Probes, Inc., Eugene, OR, Chapter 4.

17. LaRochelle, W. J. and Froehner, S. C. (1986) Determination of the tissue distributions and relative concentrations of the postsynaptic 43-kDa protein and the acetylcholine receptor in *Torpedo. J. Biol. Chem.* **261,** 5270–5274.

18. Briggs, J. and Panfili, P. R. (1991) Quantitation of DNA and protein impurities in biopharmaceuticals. *Anal. Chem.* **63,** 850–859.

19. Wong, S. S. (1991) Reactive groups of proteins and their modifying agents, in *Chemistry of Protein Conjugation and Crosslinking*, CRC, Boston, MA, pp. 27–29.

20. Kohanski, R. A. and Lane, M. D. (1985) Receptor affinity chromatography. *Ann. N Y Acad. Sci.* **447,** 373–385.

21. Orr, G. A. (1981) The use of the 2-iminobiotin–avidin interaction for the selective retrieval of labeled plasma membrane components. *J. Biol. Chem.* **256,** 761–766.

22. Hoffmann, K., Wood, S. W., Brinton, C. C., Montibeller, J. A., and Finn, F. M. (1980) Iminobiotin affinity columns and their application to retrieval of streptavidin. *Proc. Natl. Acad. Sci. U. S. A.* **77,** 4666–4668.

3

Optimization of Detection and Quantification of Proteins on Membranes in Very High and Very Low Abundance Using Avidin and Streptavidin

Sara R. Zwart and Brandon J. Lewis

Summary

Numerous methods have been published for the detection of protein using avidin–biotin technology. Complications can arise using this system when the protein of interest is in extremely high or low abundance. The ability to successfully detect high- or low-abundance proteins is dependent on the detection system selected. The expression of endogenous biotinylated proteins in rat liver homogenate will be used to illustrate the methods utilized for the detection of high- and low-abundance proteins on a membrane. The advantages and disadvantages of enzymatic and direct fluorescence detection systems are discussed.

Key Words: Biotin; avidin; detection; western blot.

1. Introduction

The high affinity of avidin for biotin (10^{-15} M) has diverse applications in many different fields for the purpose of detecting molecules. These applications include the detection of endogenous biotinylated proteins as well as the detection of proteins that have been chemically conjugated to biotin (biotin labeling) *(1–3)*. Biotin labeling of proteins is a highly effective means for detection as the conjugation of biotin to proteins allows for the assimilation of a highly specific label that generally does not perturb the function of the target protein or molecule *(4–6)*. Furthermore, there are many additional benefits to using an avidin–biotin detection system, including availability, costs, and ease of application.

From: *Methods in Molecular Biology, vol. 418: Avidin-Biotin Interactions, Methods and Applications*
Edited by: R. J. McMahon © Humana Press, Totowa, NJ

Two other forms of avidin exist with similar affinity for biotin, Streptavidin and Neutravidin. Streptavidin is produced by fermentation of *Streptomyces avidini*, whereas Neutravidin is a deglycosylated and isoelectrically neutral form of avidin. They are often used instead of avidin because of their lower extent of non-specific binding, which reduces background signals. Similar to avidin, their structures consist of four identical subunits, each containing a binding site for biotin. They can also be conjugated to an enzyme [i.e., horseradish peroxidase (HRP) or alkaline phosphatase] or a fluorescent dye to visualize biotinylated proteins on membranes. Streptavidin is stable over a wide pH range and decomposes only in the presence of SDS at temperatures above 60°C.

This chapter will focus on the application of the avidin–biotin system for the detection of biotinylated proteins in low or high abundance after separation by SDS–PAGE and transfer to polyvinyldifluoride (PVDF).

2. Materials

1. Non-fat dried milk (NFDM).
2. Tris-buffered saline, Tween (TBS-T): 20 mM Tris–HCL, pH 7.2, 150 mM NaCl, and 0.05% (v/v) Tween-20 (Sigma, St. Louis, MO).
3. Amido black stain solution: 0.2% (w/v) amido black (Napthol blue black), 40% methanol, and 10% glacial acetic acid Fisher, Pittsburgh, PA, USA.
4. Destain solution: 40% methanol and 10% glacial acetic acid.
5. Methanol.
6. Neutravidin (Pierce, Rockford, IL).
7. Neutravidin-alkaline phosphatase (Pierce).
8. Enhanced chemi-fluorescence (ECF) substrate (Amersham Pharmacia, Piscataway, NJ).
9. Alexa Fluor430 (Molecular Probes, Eugene, OR).
10. Size exclusion columns (DG-10) (Bio-Rad Laboratories, Hercules, CA).
11. Sodium bicarbonate, pH 8.3 (Fisher).
12. Phosphate-buffered saline: 20 mM sodium phosphate, pH 7.2, and 150 mM NaCl.
13. Storm 840 Optical Scanner System and ImageQuant Solutions for Windows NT software (Amersham Pharmacia Biotech).
14. Biotin (Sigma).
15. Sodium azide (Sigma).

3. Methods

The methods described below outline (i) the general method for detection of high-abundant proteins and (ii) low-abundant proteins using avidin–biotin technology. The detection of pyruvate carboxylase (PC) and acetyl CoA carboxylase (ACC) in rat liver homogenates will illustrate the methods utilized for the detection of high- and low-abundance proteins, respectively, using different avidin–biotin detection systems.

3.1. Instrumentation and Types of Detection Systems

Although high- or low-abundance proteins can ultimately be detected using various types of film or imaging systems, fluorescent or chemiluminescence detection using an optical scanner is more sensitive and linear over a wider range compared to standard film (*see* **Figs. 1** and **2**). For these reasons, we will describe methods for detection using the Storm 840 Optical Scanning System. The Storm 840 from Molecular Dynamics is an optical scanner that can detect fluorescence from storage phosphor screens or inherently fluorescent samples. The Storm 840 scanner excites fluorescent samples with a beam of light at 540 nm and detects light emitted at a wavelength of 520 nm or longer. Several approaches for the production of fluorescent molecules with these parameters have been described *(7,8)*.

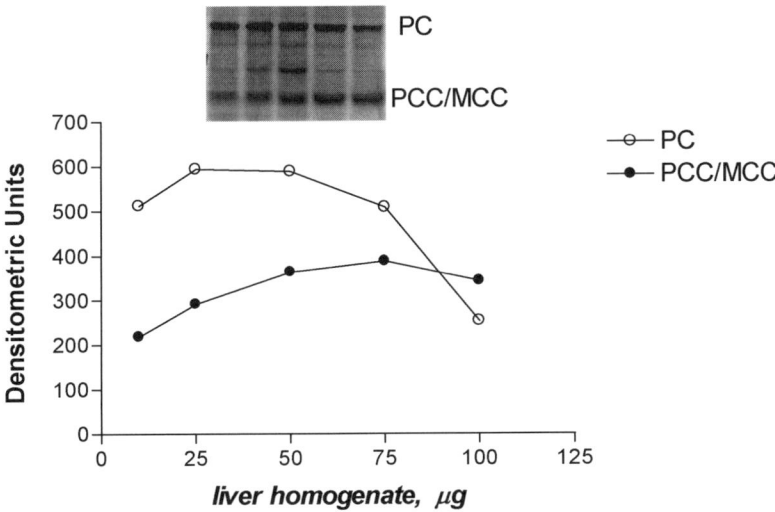

Fig. 1. Detection and quantification of biotinylated polypeptides in rat liver homogenate using film. Various amounts of rat liver homogenate were resolved by SDS–PAGE and transferred to polyvinyldifluoride (PVDF). The blots were probed with Neutravidin–horseradish peroxidase (HRP) and then visualized with enhanced chemiluminescence. The blot was exposed to film for 1 s. Open circles, pyruvate carboxylase; closed circles, propionyl CoA carboxylase and methylcrotonyl CoA carboxylase combined intensity.

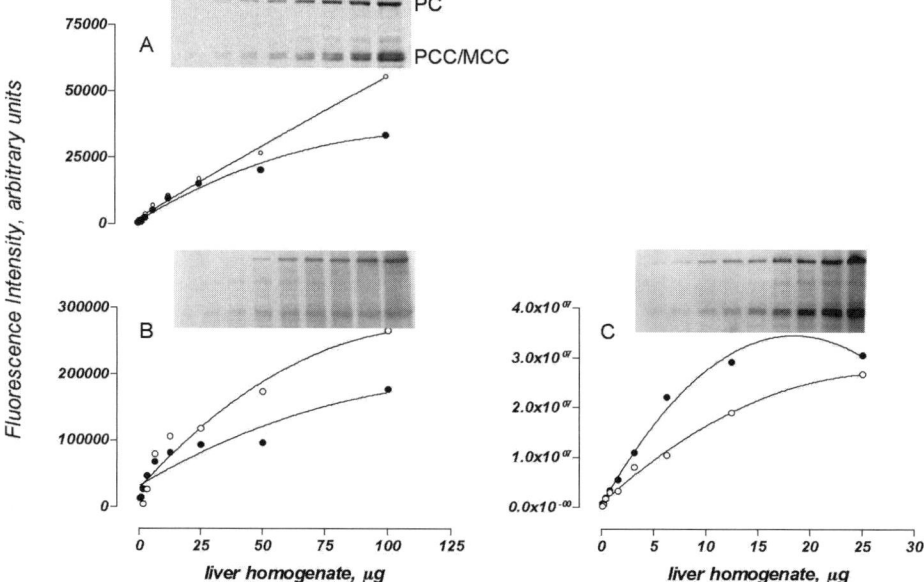

Fig. 2. Detection and quantification of biotinylated polypeptides in rat liver homogenate. Various amounts of rat liver homogenate were resolved by SDS–PAGE and transferred to polyvinyldifluoride (PVDF). The blots were probed with (**A**) avidin–Alexa Fluor430 conjugate; (**B**), avidin–horseradish peroxide (HRP) and ECL; and (**C**), avidin–AP and ECF. Open circles, pyruvate carboxylase; closed circles, propionyl CoA carboxylase and methylcrotonyl CoA carboxylase combined intensity [Reprinted from (**9**), with permission from Elsevier].

3.1.1. Enzyme Systems

When visualizing low-abundance proteins, enzymatic detection systems have advantages over fluorescent systems. The signal of the protein to be detected is amplified by the enzymatic reaction; therefore, low-abundance proteins can readily be detected. There are disadvantages, however, in that there is usually a higher background signal, and streaking or smearing can appear around the protein of interest if care is not taken when placing the membrane on the scanner.

ACC is an endogenous biotinylated protein that is present in low abundance in rat liver. Neutravidin conjugated to alkaline phosphatase (avidin-alkaline phosphatase) generates a fluorescent product when incubated with fluorescent substrate that is detectable by the STORM 840 and will be utilized to illustrate the detection of ACC. This substrate, under the commercial name of ECF western blotting substrate, exhibits a maximum excitation wavelength of

440 nm and a maximum emission of 560 nm. Alternatively, ECL-Plus can be used as a western blotting substrate with avidin conjugated to HRP (avidin–HRP). ECL-Plus produces a chemiluminescent product that also exhibits fluorescent properties. The maximum excitation wavelength of ECL-Plus is 420 nm, and the maximum emission wavelength is ~460 nm. Although this emission spectrum is less than the optimal detection spectrum of the Storm 840, the system is sensitive enough to detect 0.004 ng biotinylated bovine serum albumin on PVDF; however, validation studies show that ECF is more sensitive than ECL-Plus (*9*).

3.1.1.1. Western blotting

As this chapter primarily focuses on the detection of proteins, discussions regarding the separation of proteins and electrophoretic transfer of proteins to PVDF will not be mentioned in detail. For the methods we used to resolve and transfer ACC and PC (*see* **Note 1**).

1. Cut the PVDF membrane to include the biotinylated protein(s) of interest (*see* **Note 2**).
2. Incubate the PVDF blot in 0.5% NFDM in TBS-T for blocking purposes for 10 min. This step prevents non-specific binding of the avidin-alkaline phosphatase to the PVDF or other proteins on the PVDF, thereby reducing background signals.
3. Incubate the blot in 0.5% NFDM in TBS-T and a 1:6000 dilution of Neutravidin avidin-alkaline phosphatase conjugate for 40 min on an orbital shaker (*see* **Note 3**).
4. Wash the membrane three times in TBS-T without NFDM, with 5 minutes of incubation on an orbital shaker between each wash.
5. After dumping off the contents of the last wash, add a chemiluminescent substrate that exhibits fluorescent properties (ECF) to the blot and incubate for 5 min (*see* **Note 4**).
6. Place blot face down on the glass surface of the Storm 840 Optical Scanner, and scan according to the manufacturer's instructions.

Although ECF detection works well for low-abundance proteins, we found that high-abundance proteins detected with ECF results in some diffusion of the fluorescent product. Likewise, high-abundance proteins detected with ECL-Plus also exhibited some complications. The fluorescent ECL-Plus product is not soluble, so there are fewer problems with product diffusion; however, high-abundance protein bands can appear with a "ghosted" appearance in the center of the band, indicating that there is a loss of signal in the interior of large dense bands. Simply loading less protein mass can generate a more linear response while using the ECF or ECL-Plus detection system. We found ECF detection to be more sensitive and linear than ECL-Plus, fluorescent detection, or standard film (**Figs. 1** and **2**).

3.1.1.2. Quantification of Data

1. Analyze the blots using ImageQuant software. A local average background correction should be used, and then the fluorescent intensity can be integrated over each protein band area.
2. The integrated area will generate a number, in arbitrary units. For quantitative results, a standard curve of known concentrations of the protein of interest should be resolved by SDS–PAGE and detected. The arbitrary number generated from the protein of interest should be within the linear part of the standard curve to ensure accurate quantitative data. If the arbitrary number is above or below the standard curve, then either a dilution or concentration of the protein should be done prior to resolving on an SDS–PAGE gel.

3.1.2. Fluorescent Dyes

Direct fluorescence detection has advantages over enzyme systems in that background noise on the blot is minimal. As this detection method does not amplify the signal of the protein to be detected, it can generally only be used for proteins that have been purified and concentrated or that are naturally present in high abundance.

This section will describe the detection of PC using fluorescent avidin detection. PC is an endogenous biotinylated protein (MW = 130 kDa) that is naturally present in high abundance in the liver. The detection system utilizes Neutravidin that has been covalently labeled with the fluorescent dye Alexa Fluor430. Alexa Fluor430 exhibits a maximum excitation wavelength of 430 nm and a maximum emission wavelength of 540 nm. The utilization of a fluorescent dye such as this has the advantage of not relying on the enzymatic activity of either alkaline phosphatase or HRP and the associated kinetic characteristics that could confound the analysis. A validation study using this method showed that the response of the fluorescent avidin conjugate was very similar to the theoretical response of serial dilution and that the fluorescence emitted and the amount of the conjugate bound to the biotinylated protein is stoichiometric (*9*).

3.1.2.1. Synthesis of Avidin–Alexa Fluor430

1. Dissolve 10 mg Neutravidin in 1 ml 0.1 mM sodium bicarbonate, pH 8.3.
2. Dissolve 5 mg of the succinimidyl ester form of Alexa Fluor 430 in 500 µl dimethyl sulfoxide and vortex slowly (this step must be done immediately before conjugation).
3. Add 400 µl of the dye solution to 1 ml of the avidin solution and protect from light while vortexing for 1 h.
4. Remove the unconjugated dye by pouring the contents of the reaction over a DG-10 column (Bio-Rad) equilibrated in PBS. Collect equal volume fractions (1 ml) and combine the peaks with the highest absorbance at 280 nm. Add sodium azide (0.2 g/l) to the sample for preservation and store the conjugate at 4°C protected from light.

3.1.2.2. WESTERN BLOTTING

1. Follow the steps outlined above for the detection of ACC using an enzymatic detection system. In step 3, add 0.5% NFDM solution in TBS-T and the avidin–Alexa Fluor430 conjugate at a concentration of 2.4 μg/ml.
2. Allow the blot to incubate with the solution for 45 min at room temperature on an orbital shaker.
3. Wash the blot three times with TBS-T for 5 min each.
4. After the third wash, place the blot face down on the Storm 840 Optical Scanner and scan according to the manufacturer's instructions.
5. Analyze the blots using ImageQuant software. A local average background correction should be used and then the fluorescent intensity can be integrated over each area.

3.2. Determining Linearity of the Detection Method

To determine the linear range of detection, a range of protein masses should be resolved on a SDS–PAGE gel and detected with an enzymatic detection system or avidin AlexaFluor430. The specificity of the detection should be confirmed through competition with excess biotin.

3.3. Testing for Specificity of the Detection Method

Endogenous and chemically biotinylated proteins can both be tested for specificity by competing the avidin detection system with excess biotin. In this example, the specificity of the detection of endogenous biotinylated proteins in rat liver homogenate was tested using avidin–Alexa Fluor430.

1. Resolve the protein mixture in at least two different lanes on an SDS–PAGE and transfer to PVDF.
2. For the competition, combine a 1:750 dilution of avidin–AlexaFluor430 in 0.5% NFDM with 0.1 μM biotin.
3. Incubate the solution on an orbital shaker for 1–2 h at room temperature to allow for complete binding.
4. Cut the two identical protein lanes out of the PVDF membrane and place into separate containers.
5. Incubate the PVDF blots in 0.5% NFDM in TBS-T for blocking purposes for 10 min.
6. Incubate one of the PVDF blots in the avidin–biotin competition mixture for 45 min and incubate the other blot in 1:750 avidin–AlexaFluor430 diluted in 0.5% NFDM.
7. Wash the blots three times with TBS-T for 5 min each.
8. Place the blots face down next to each other on the Storm 840 scanner and scan according to the manufacturer's instructions. Results for testing the specificity of endogenous biotinylated proteins in rat liver homogenate are presented in **Fig. 3**.

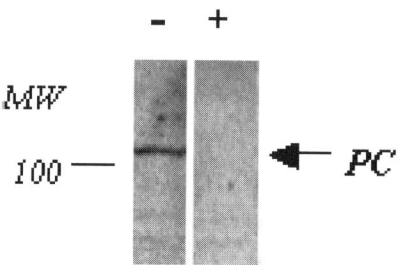

Fig. 3. Avidin competition.

The abundance of a particular protein must be considered when deciding which detection method to use. High-abundance proteins can easily be detected using an enzymatic or direct fluorescence method, depending on the mass of the protein resolved on the membrane. Endogenous biotinylated proteins must be also carefully considered when using crude biological samples, as high-abundance endogenous biotinylated proteins (e.g., PC) can compete with a lower abundance protein for a particular substrate and leave the protein of interest undetectable *(10)*. As illustrated here, careful attention to loading parameters can produce highly sensitive and very accurate quantification.

4. Notes

1. The proteins of interest (0.1 mg liver homogenate protein for the analysis of ACC, 0.08 mg for PC) were resolved by SDS–PAGE. ACC was separated using a 3% acrylamide stacking gel (0.125 M Tris HCl, pH 6.8, 0.1% SDS, 3% acrylamide, 0.05% APS, and 30 µl Tetramethylethylenediamine (TEMED)) and a 5% separating gel [0.375 M Tris–HCl, pH 8.8, 0.1% SDS (Fisher), 5% acrylamide (1:29 bisacrylamide : acrylamide), 0.05% ammonium persulfate (Sigma), and 30 µl TEMED (Fisher)]. PC was separated similarly to ACC except an 8% gel with separating gel buffer of pH 8.0 was used. Sample homogenates (75 µg total protein) were diluted into 20 µl of sample dilution buffer [0.025 M Tris-base, 0.23% SDS, 35% (v/v) glycerol, 0.035 mg/ml bromophenol blue, and 1.43 mmol/l β-mercaptoethanol (Sigma)] before loading onto the stacking gel *(11)*. Proteins were electrophoresed overnight with running buffer [0.025 M Tris base, 0.2 M glycine (Sigma), and 0.1% SDS] at 47 V until the dye front was ∼1 cm from the bottom of the gel.

   The resolved gel was equilibrated in cold transfer buffer [0.03 mol/l Tris-base, 0.2 mol/l glycine, and 60% (v/v) methanol] for 10 min and was then electroblotted to PVDF (Immobilon-P, Millipore, Bedford, MA) for 2 h at 12 V. The blot was stained with amido black stain [50% (v/v) methanol, 10% (v/v) glacial acetic acid, and naphthol blue black] and destained [50% (v/v) methanol and 10% (v/v) glacial acetic acid] and then washed with two changes of methanol before allowing it to air dry.

2. Cutting the PVDF membrane must be done carefully because there is a risk of cutting off or cutting through the protein of interest. Molecular weight standards should always be resolved concurrently with the protein of interest for this reason. It is ideal to cut as close to the protein of interest as possible without cutting it off, as any other endogenous protein or high-abundance protein could compete for the substrate. If the protein of interest is not purified, then one must consider that the endogenous biotinylated proteins will also be detected with an avidin-detection system. If the protein of interest is not purified and is close in molecular weight to one of the endogenous biotinylated proteins, then purification or a different detection system should be considered.

   Also, when cutting the PVDF membrane, wear gloves when touching the membrane, and do not touch forceps or scissors to the blot anywhere other than the edge of the blot. If forceps touch the blot, a mark or streak may appear in the final scanned image.

3. The dilution of avidin in NFDM should be optimized each time new avidin is purchased or produced. The NFDM solution should just barely cover the blot and should flow freely over the blot when rotated on an orbital shaker. If the volume of NFDM covering the blot is too high, this will not allow for maximum movement of the avidin conjugate over the blot. Also, if the blot begins to float on top of the NFDM solution, the container must be shaken periodically during the incubation to ensure that the blot does not dry out. Alternatively, the blot can be turned over and allowed to float on top of the NFDM solution if it begins to float.

4. For best results, apply the ECF substrate to a glass or plastic plate and place the blot down (protein side facing down). Once the blot is placed, take care not to disturb or move the blot because this will cause smearing once the blot is scanned. Incubate for 5 min. In one motion, carefully pick up the blot and quickly wash in TBS-T (the product of the enzymatic reaction is soluble, so do not wash more than 1–2 s). The wash will remove some of the background.

Acknowledgments

The authors thank Dr. Robert J. McMahon for his support, advice, and encouragement of the project.

References

1. Neumaier, M., Fenger, U., and Wagener, C. (1986) *Anal Biochem* **156**, 76–80.
2. Dunn, M. J. (1994) *Methods Mol Biol* **32**, 227–32.
3. Ruggiero, F. P., and Sheffield, J. B. (1998) *J Histochem Cytochem* **46**, 177–83.
4. Magnusson, S., Hou, M., Hallberg, E. C., Breimer, M. E., and Wadenvik, H. (1998) *Thromb Res* **89**, 53–8.
5. Goldman, E. R., Balighian, E. D., Mattoussi, H., Kuno, M. K., Mauro, J. M., Tran, P. T., and Anderson, G. P. (2002) *J Am Chem Soc* **124**, 6378–82.

6. Khosravi, M. J., and Morton, R. C. (1991) *Clin Chem* **37,** 58–63.
7. Hitt, A. L., Laing, S. D., and Olson, S. (2002) *Anal Biochem* **310,** 67–71.
8. Martin, K., Hart, C., Schulenberg, B., Jones, L., and Patton, W. F. (2002) *Proteomics* **2,** 499–512.
9. Lewis, B., Rathman, S., and McMahon, R. J. (2003) *J Nutr Biochem* **14,** 196–202.
10. Praul, C. A., Brubaker, K. D., Leach, R. M., and Gay, C. V. (1998) *Biochem Biophys Res Commun* **247,** 312–4.
11. Lewis, B., Rathman, S., and McMahon, R. (2001) *J Nutr* **131,** 2310–5.

4

Use of the Avidin (Imino)Biotin System as a General Approach to Affinity Precipitation

Ruth Freitag and Frank Hilbrig

Summary

Biospecific interactions are used in many capturing and bioseparation steps. A typical situation is the coupling of a biospecific ligand to a chromatographic stationary phase for affinity chromatography. This approach has two possible drawbacks. The first is that a chromatographic column may be awkward to use in experimental setups; the second is related to the need to develop a dedicated coupling chemistry for any given affinity ligand. In affinity precipitation, the biospecific affinity ligand is instead linked to a stimuli-responsive molecule to yield a so-called affinity macroligand (AML). Upon stimulation, such molecules show abrupt yet reversible precipitation from aqueous solution. Capture by affinity precipitation just requires the addition of the stimuli-responsive AML to the raw target solution followed by selective precipitation of the formed affinity complex via the application of the stimulus. The need for the synthesis of a dedicated AML may be circumvented by the use of an avidin-activated stimuli-responsive precursor, to which any biotinylated affinity ligand can be securely linked via the well-known strong avidin–biotin interaction.

Key Words: Affinity chromatography; affinity precipitation; avidin; bioconjugate; biotin; LCST, peptide ligands; PNIPAAm, protein purification; stimuli-responsive materials.

1. Introduction

The interaction between the egg white protein avidin and biotin (vitamin H) although nominally 'non-covalent' is so strong that for all practical purposes a permanent complex is formed when the two molecules come into contact *(1)*. When iminobiotin is used instead of biotin (*see* **Scheme 1** for the two

From: *Methods in Molecular Biology, vol. 418: Avidin-Biotin Interactions, Methods and Applications*
Edited by: R. J. McMahon © Humana Press, Totowa, NJ

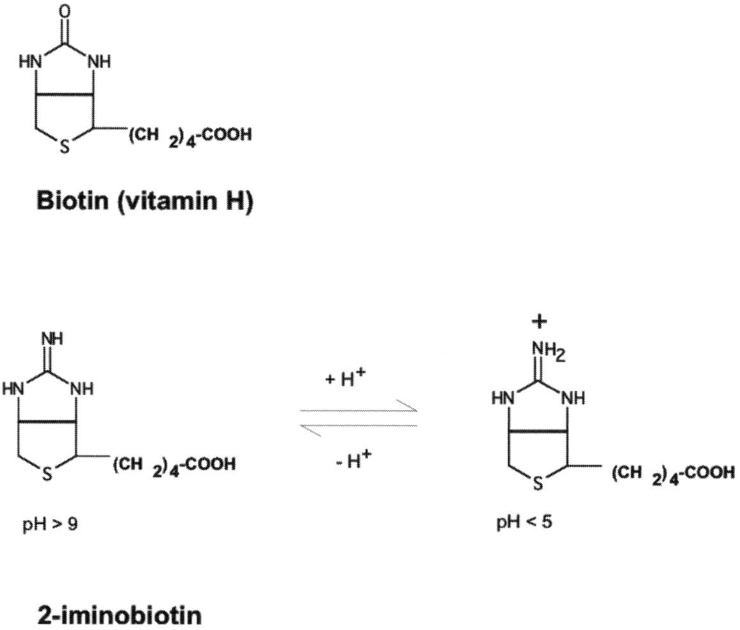

Biotin (vitamin H)

2-iminobiotin

Scheme 1. Structures of biotin (top) and iminobiotin (bottom).

structures) the complex stability is pH dependent and dissociation is possible by lowering the pH to 4. Streptavidin can be used as alternative to avidin in both cases. The strong binding of avidin to (imino)biotin has found several applications in bioseparation, when biospecific interactions are required. The simplest application would be the isolation of avidin using biotin as affinity ligand or vice versa.

More importantly, however, the system has established itself as a simple generic means to produce affinity supports. A host of 'biotinylated', biologically interactive molecules can be obtained from dedicated suppliers of chemicals and biologicals. Concomitantly, it is relatively straightforward to develop a protocol for linking avidin to a surface, a stimuli-responsive molecule, a hydrogel, or a three-dimensional scaffold. By snapping the two building blocks (avidin-activated support and biotinylated affinity ligand) together, a wide variety of specifically interactive structures can be produced. Alternatively, a fusion protein carrying the biotin-binding domain of avidin can be linked as interactive ligand to a biotin-activated material.

One area that may benefit from such a general approach is affinity precipitation *(2)*. In affinity precipitation, an affinity ligand capable of recognizing and binding ('capturing') the target molecule is linked to a stimuli-responsive molecule such as the thermo-responsive poly(N-isopropylacrylamide) (PNIPAAm) *(3,4)* to yield a stimuli-responsive affinity macroligand (AML). PNIPAAm dissolves well in cold water, yet becomes insoluble if a certain 'critical' temperature is surpassed (34°C, in the case of PNIPAAm dissolved in pure water). The effect, which manifests itself by the rapid occurrence of a precipitate or—at lower concentration—a clouding of the solution, is fully reversible, and the precipitate redissolves readily when the solution's temperature is lowered below the critical one.

Such a lower critical solution temperature (LCST) is typically observed in systems that combine a negative dissolution enthalpy with a negative dissolution entropy, for example, certain macromolecules that contain both hydrophobic and hydrophilic units, yet which by their structure are prevented from forming micelles. As the temperature increases, the strength of the dissolution-aiding H-bridges decreases, while the effect of the negative dissolution entropy increases. At a certain temperature, the Gibbs free energy term for dissolution becomes positive and phase separation/precipitation occurs. The particular critical temperature of a given system is a function of the solution's composition. For example, most salts lower the phase separation temperature, whereas other molecules have been described that increase it *(5,6)*.

In affinity precipitation, stimuli-responsive bioconjugates (AMLs) are used to capture the target molecule in homogeneous solution *(7–9)*. Then, the complex is selectively precipitated by the application of the stimulus and removed from the supernatant. The target molecule is then released either directly from the precipitate or after redissolution of the complex in cold dissociation buffer. Affinity precipitation has a wide application spectrum from large-scale product capture in biotechnology to analytical high-throughput screening, where the ability to dissolve and precipitate the capturing agent as desired presents a considerable advantage. Given the known difficulty in creating bioconjugates from the individual building blocks, the alternative of using a stimuli-responsive PNIPAAm–avidin conjugate together with a biotinylated commercial affinity ligand to create the AML should be considered as a quick and reliable alternative.

2. Materials

2.1. Synthesis of the Biotin-Based Chain Transfer Agent

1. Chemicals: Swollen cysteamine 4-methoxytrityl resin (Calbiochem-Novabiochem, USA, as used in solid-phase peptide synthesis), biotin (USP grade), O-(benzotriazol-1-yl)-N,N,N´,N´-tetramethyluronium hexa-fluorophosphate (HBTU, 97%), N,N-dimethylformamide (DMF, grade: 'for peptide synthesis'), dichloromethane (anhydrous), trifluoroacetic acid (TFA, puriss grade), triisopropylsilane (TIS, 98%), 4-methylmorpholine (NMM, 99.5%), toluene (anhydrous). Unless specifics are given, chemicals can be obtained from any reliable chemical supplier.

2. Special equipment: 50-mL reaction vessel (glass) with fritted bottom, as used in solid-phase peptide synthesis.

2.2. Chain Transfer Polymerization (Telomerization) of NIPAAm

1. Chemicals: 3-mercaptopropionic acid (MPA, chain transfer agent, *see* **Note 1**), N-isopropylacrylamide (NIPAAm, monomer, should not contain any inhibitor, *see* **Note 2**), 2,2′-azoisobutyronitrile (AIBN, radical chain starter), methanol (<0.02 % water, solvent), diethylether, n-hexane, acetone.
2. AIBN must be recrystallized in diethylether prior to use and can then be stored for several days at –20°C in the dark.

2.3. Synthesis of PNIPAAm-Avidin

1. Chemicals: N-hydroxysuccinimide (NHS), N,N′-dicyclohexylcarbodiimide (DCC), dichloromethane (anhydrous), DMF (anhydrous), PNIPAAm-COOH (prepared according to methods described in **Subheading 3.2.** or obtained from polyTag Technology AG, Männedorf, Switzerland).
2. Solution of avidin (from egg white, affinity purified) in 0.2 M borate buffer, pH 8.6.
3. Saturated ammonium sulphate solution.

2.4. Synthesis of PNIPAAm-Iminobiotin

1. 2-Iminobiotin NHS ester, DMF (anhydrous), PNIPAAm-NH$_2$ (produced according to methods described in **Subheading 3.2.** or obtained from polyTag Technology AG)
2. 0.1 M Borate buffer, pH 8.75.
3. Saturated ammonium sulphate solution.
4. 0.01 M Triethylamine (TEA) buffer, pH 8.
5. 0.1 M Carbonate buffer, pH 10.8.

2.5. Coupling of Biotinylated Affinity Ligands to PNIPAAm-Avidin

1. PNIPAAm-avidin (prepared according to methods described in **Subheading 3.3.** or obtained from polyTag Technology AG)
2. The desired affinity ligand bearing a biotin group.
3. 0.5 M Borate buffer, pH 9.
4. Saturated ammonium sulphate solution.

2.6. Affinity Precipitation

1. Binding buffer: defined by the system; in the case of capture of avidin by PNIPAAm-iminobiotin: 0.1 M sodium carbonate buffer, pH 10.8.

2. Dissociation buffer: defined by the system. In the case of avidin capture by PNIPAAm-iminobiotin: 0.1 M ammonium acetate buffer, pH 4.0, containing 0.5 M NaCl.

2.7. HABA Assay (see ref. 10)

1. Biotin solution (2 mM) in water.
2. 4-Hydroxyazobenzene-2´-carboxylic acid (HABA) (10 mM) solution in water.

2.8. Determination of the Average Molecular Weight of PNIPAAm-COOH by End Group Titration

1. NaOH, 0.1 M, phenolphthalein (indicator).

3. Methods

Access to agents for affinity precipitation has been facilitated considerably over the last 5 years. In the past, the only possibility of obtaining suitable bioconjugates was an in-house synthesis of the required agent. Today, stimuli-responsive bioconjugates bearing generically applicable ligands such as avidin, (imino)biotin, Protein A, or chelated Ni ions are commercially available from polyTag Technology AG (http://www.polytag.ch). These agents and in particular the avidin/(imino)biotin-activated ones may be used directly according to the protocols described below. In addition, proven and easy to follow synthesis protocols are given in this section that will allow any person with a reasonable background in chemical synthesis to prepare the required agents.

3.1. Synthesis of the Biotin-Based Chain Transfer Agent

The synthesis of PNIPAAm molecules bearing a biotin end group requires a biotin derivative as chain transfer agent. 2-Biotinamidoethanethiol can fulfil this purpose. For the synthesis of this molecule, proceed as follows.

1. Secure the 50-mL reaction vessel with fritted bottom in the hood.
2. Inside the flask place 500 mg of the swollen cysteamine 4-methoxytrityl resin and 15 mL of a solution containing both 268 mg biotin and 521 mg HBTU in DMF.
3. Stir by nitrogen flow bubbling through the fritted bottom for 4 h at room temperature.
4. Remove the liquid by applying a vacuum to the fritted bottom and wash the resin well first with DMF and then with dichloromethane.
5. Add 20 mL of a solution containing 5% (v/v) of TFA, 5% (v/v) of TIS, and 241 μL NMM in dichloromethane and stir again by nitrogen bubbling for 2 h at room temperature.
6. Recover the liquid by applying a vacuum to the fritted bottom.

7. Pour the liquid into a 200-mL glass flask containing 50 mL toluene.
8. Evaporate the solvent in a vacuum rotary evaporator.
9. From this procedure one should obtain approximately 200 mg of white powder with the following characteristics (*see* **Note 3**)—^{1}H-NMR (DMSO): 0.7–1.4 (m, 6H, CH–(CH$_2$)–(CH$_2$)$_3$–CON), 1.6–1.8 (m, 1H, (–CH$_2$–SH), 2.0 (m, 1H, –CH–HCH–S–), 2.1–2.3 (m, 4H, –CH$_2$–CON–CH$_2$–), 2.4–2.6 (m, 1H, –CH–HCH–S–), 2.7–3.0 (m, 3H, –CH$_2$–SH + –S(CH)–CH–CH$_2$–), 3.8 (m, 1H, –CH–CH–NH(CH)–), 4.0 (m, 1H, –CH$_2$–CH–NH(CH)–), 5.8–6.3 (m, 3H, –CO–NH); ESI-MS m/z: 304.113 (H$^+$).

3.2. Chain Transfer Polymerization (Telomerization) of NIPAAm

For the polymerization of approximately 4 g of PNIPAAm with a terminal carboxylic acid end group (PNIPAAm-COOH, suitable for linkage of avidin, *see* **Subheading 3.3.**) proceed as follows (for the production of PNIPAAm-NH$_2$ or PNIPAAm-biotin, *see* **Note 1**).

1. In a 200-mL one-necked glass flask place 5 g of NIPAAm (monomer) in 50 mL methanol.
2. Add 0.006 mol equivalent of AIBN (radical starter) and 0.06 mol equivalent of MPA (chain transfer agent).
3. Replace the air in the flask by argon and keep the reaction mixture under argon atmosphere throughout the reaction.
4. Heat the well-stirred mixture to 65°C under reflux and allow the reaction to proceed for 3 h.
5. Remove the methanol from the reaction mixture in a vacuum rotary evaporator.
6. Dissolve the resulting white product in 45 mL acetone and pour it slowly (dropwise) under vigorous stirring into 500 mL hexane (*see* **Note 4**).
7. The produced PNIPAAm-COOH should precipitate in the form of a fine white solid that can be recovered by vacuum filtration in a Buchner funnel.
8. Dry the polymer to constant weight in a vacuum oven at 40°C.
9. The average of the molecular weight and the weight distribution of the produced polymer should be determined at this point by end-group titration (*see* **Subheading 3.9.**) and MALDI-TOF mass spectrometry (MS) (*see* **Subheading 3.8.**) (*see* **Note 5**). Various validated service suppliers exist that provide such mass spectra on a commercial basis.

3.3. Synthesis of PNIPAAm-Avidin

Avidin is linked to PNIPAAm-COOH by carbodiimide coupling as follows.

1. Dissolve 1 g PNIPAAm-COOH in 20 mL anhydrous dichloromethane under a moisture-free argon atmosphere and cool to 4°C.
2. Add 2 mol equivalent NHS and 2 mol equivalent DCC under constant stirring and continue to stir for another 4 h always maintaining the mixture at 4°C.

3. Allow the mixture to warm to room temperature. An insoluble precipitate (dicyclohexyl urea) will form, which should be removed by filtration.
4. Then the product (PNIPAAm-NHS) is recovered from the reaction mixture by precipitation from anhydrous diethyl ether.
5. The fine precipitate is filtered and dried to constant weight in a vacuum oven.
6. For coupling of avidin, 400 mg PNIPAAm-NHS is dissolved in 1 mL anhydrous DMF and added incrementally at 4°C to a gently stirred solution of 120 mg avidin in 5 mL 0.2 M borate buffer (pH 8.6) (*see* **Note 6**).
7. Continue stirring for another 2 h at 4°C and allow the mixture to warm to room temperature.
8. Clarify the solution by centrifugation (10,000 g for 10 min).
9. Add 600 μL saturated ammonium sulphate solution (*see* **Note 7**).
10. Purify the PNIPAAm–avidin bioconjugate by three thermoprecipitation cycles. A thermoprecipitation cycle calls for precipitation at 30°C and recovery of the precipitate via centrifugation (10,000 g,10 min, 30°C) followed by redissolution of the recovered pellet in fresh borate buffer (do not forget to add the saturated ammonium sulphate solution) at 4°C.
11. Lyophilize for storage.
12. After lyophilization, the bioconjugate should be stable for at least 12 months if kept in the dark at 4°C.

3.4. Synthesis of PNIPAAm-Iminobiotin

Iminobiotin is linked to PNIPAAm-NH$_2$ as follows (*see* **Note 8**).

1. Dissolve 1 g PNIPAAm-NH$_2$ in 50 mL 0.1 M borate buffer, pH 8.75, and cool to 4°C.
2. Dissolve 60 mg 2-iminobiotin NHS ester in 2 mL anhydrous DMF (final concentration 3×10^{-5} M).
3. Add the 2-iminobiotin NHS ester solution dropwise to the gently stirred PNIPAAm-NH$_2$ solution maintaining 4°C throughout.
4. Keep stirring for 2 h and allow the mixture to warm to room temperature.
5. Add 1 mL saturated ammonium sulphate solution (*see* **Note 7**).
6. Remove the product from the reaction mixture by thermoprecipitation (precipitation at 30°C and recovery of the precipitate via centrifugation (10,000 g,10 min, 30°C).
7. Dissolve the precipitate in 10 mL 0.01 M TEA buffer to convert the precursor (PNIPAAm-iminobiotin HBr) into the final product.
8. Recover the final product by thermoprecipitation.
9. Dissolve the precipitate in distilled water and lyophilize for storage.

3.5. Coupling of Biotinylated Affinity Ligands to PNIPAAm-Avidin

Any biotinylated ligand may be linked to PNIPAAm-avidin as follows (*see* **Note 9**).

1. Dissolve the biotinylated ligand (1 mg/mL) in 0.5 M borate buffer (pH 9).

2. Add the solid PNIPAAm-avidin directly to this solution (molar ratio 1:1) and stir gently for 2 h at 18°C.
3. Add 10 % (v/v) of the saturated ammonium sulphate solution (*see* **Note 7**).
4. Purify the AML by repeated (3×) thermoprecipitation (30°C) and centrifugation at 30°C and 10,000 g (10 min) followed by redissolution in a fresh batch of buffer/ammonium sulphate solution at 4°C.
5. Depending on the nature of the molecule used as affinity ligand, the final precipitate may be redissolved in pure water and lyophilized for storage.

3.6. Affinity Precipitation

The exact conditions for the recovery of a given target molecular by affinity precipitation depend on the nature of the employed interaction. Here, a general protocol is given with the particular conditions for the recovery of avidin by PNIPAAm-iminobiotin indicated as example.

1. Choose a suitable binding buffer that supports and association buffer that interrupts the intended affinity interaction well (*see* **Note 10**). Indications can usually be taken from the corresponding affinity chromatography protocols. In the case of avidin recovery by PNIPAAm-iminobiotin, capture can, for example, take place directly in the cell-free supernatant even in the presence of serum.
2. Stir the AML into the target molecule solution using a 10-fold molar excess (*see* **Note 11**). The temperature should be at least 5°C below the critical solution temperature of the AML in the binding environment during this process ('binding temperature', *see* **Note 12**).
3. Stir at binding temperature for 30 min (*see* **Note 13**).
4. Raise the temperature at least 5°C above the critical solution temperature of the AML in the binding environment ('precipitation temperature', *see* **Note 12**) and remove the precipitated affinity complex by centrifugation (10,000 g, precipitation temperature,10 min).
5. Remove entrapped impurities by repeated thermoprecipitation in binding buffer [redissolution of the precipitate in 4°C cold binding buffer (*see* **Note 14**), precipitation at precipitation temperature, recovery of the precipitate at 10,000 g for 10 min at precipitation temperature, followed by redissolution of the pellet in 4°C cold binding buffer, *see* **Note 15**].
6. For release of the target molecule, redissolve the pellet in 4°C cold dissociation buffer (*see* **Note 15**). In the case of avidin capture by PNIPAAm-iminobiotin, 0.1 M ammonium acetate buffer, pH 4.0, containing 0.5 M NaCl can be used as dissociation buffer.
7. Remove the AML from the purified product by thermoprecipitation followed by centrifugation (*see* **Notes 16** and **17**).

3.7. HABA Assay

The amount of avidin in the dissolved AML can be determined by Green's HABA assay *(10)* as follows.

1. Prepare a 2 mL solution of the AML containing approximately 0.2–0.5 mg of avidin.
2. Place an aliquot of the solution in a 1-cm quartz cuvette.
3. Use an appropriate solution as reference, for example, a solution containing PNIPAAm instead of PNIPAAm-avidin. Place this solution in the reference cuvette.
4. At 280 nm, determine the absorption (A) of the avidin-containing solution in comparison to the reference solution.
5. Calculate the amount of avidin (g in mg) contained in this 2 mL solution using an extinction coefficient (ε) of 106,081 (mol/cm) and a molecular weight (M) of avidin of 67,500 g/mol (*see* **Note 18**) according to the formula $\dfrac{2 \times A \times M}{\varepsilon} = g$.
6. Add 50 μL of 10 mM HABA solution to the avidin solution and determine the absorption (A_1) at 500 nm when the value is stable for at least 15 s.
7. Add 50 μL of 2 mM biotin solution to this solution and determine the absorption (A_2) at 500 nm when the value is stable for at least 15 s.
8. Calculate the number of biotin binding places per avidin (BP) according to the formula $BP = \dfrac{(A_1 - A_2) \times M}{17,000 \times g}$.

3.8. Determination of the Average Molecular Weight of PNIPAAm-COOH by End Group Titration

1. The (number) average molecular weight of PNIPAAm-COOH can be determined by titration with 0.1 M NaOH using phenolphthalein as indicator (*see* **Note 19**).
2. The (number) average molecular weight of the polymer is calculated as the absolute amount (mass) of polymer in the sample divided by the number of COOH groups determined by titration.

3.9. Characterization of the Polymer Polydispersity by MALDI-TOF MS

The molecular weight and weight distribution of PNIPAAm is easy to analyse by MALDI-TOF MS (*see* **Note 20**). If no MALDI-TOF MS is available in the group, outsourcing should be considered. MALDI-TOF analysis is available from a number of specialized centres inside and outside academia. The sample concentration should be 5 mg/mL in pure water, sinapinic acid (3,5-dimethoxy-4-hydrocinnamic acid) in 30% CH_3CN containing 0.1% trifluoric acid (TFA) may be used as matrix. For a characterization of the molecular weight distribution and the determination of both the number and the mass average of the molecular weight proceed as follows.

1. You will obtain a molecular weight distribution in the form of peaks, each representing a given degree of polymerization as shown in **Fig. 1**. The height of the peak represents the abundance of this particular type of molecule in the preparation, the molecular weight of this particular species can be taken directly

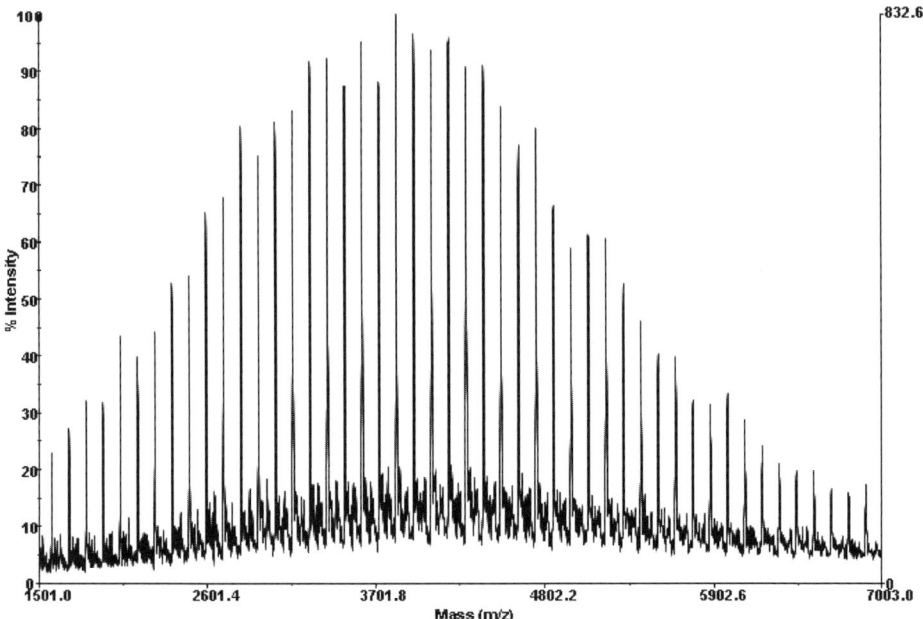

Fig. 1. MALDI-TOF mass spectrum of PNIPAAm.

form the x axis. In the case of PNIPAAm, the distance between the peaks should
be 113 g/mol corresponding to one monomeric unit.

2. The number average M_n of the molecular weight is calculated from the spectra
   according to the formula $M_n = \dfrac{\sum N_i M_i}{\sum N_i}$.

3. The mass average M_w of the molecular weight is calculated from the spectra
   according to the formula $M_w = \dfrac{\sum N_i M_i^2}{\sum N_i M_i}$.
   The polydispersity ('broadness') of the molecular weight distribution can be
   estimated as $P = M_w/M_n$.

3.10. Turbidity Curve Measurements

The critical solution temperature (LCST) of the stimuli-responsive bioconjugate is defined as the inflection point of its optical density versus temperature or turbidity curve. Because PNIPAAm has a very sharp phase transition,
the temperature where the optical density reaches half height can be used as
reasonable approximation of the LCST.

1. Measurements are done at 500 nm using a Lambda 20 spectrometer (Perkin-
   Elmer, Norwalk, CT, USA) equipped with a PTP 1 thermostat and a temperature
   sensor directly inserted into the reference cell (*see* **Note 21**).

2. Pipette a solution of 1% (w/w) of PNIPAAm (or its derivative) dissolved in the solution of interest into the measurement cuvette (*see* **Note 22**).

3. Pipette the same solution of interest without the PNIPAAm into the reference cuvette.

4. Measure the turbidity curve at a heating rate of 0.5°C/min (*see* **Note 23**).

5. Record also the turbidity curve as the solution cools down (*see* **Note 24**).

4. Notes

1. A chain transfer agent is a necessary part of any chain transfer polymerization mixture. During the reaction it is incorporated into each polymer chain as an end group, and its function is to reduce both the average molecular weight and the broadness of the molecular weight distribution in the polymer preparation. Nearly any small molecule containing a thiol group can be used as chain transfer agent. As the chain transfer agent is incorporated into the polymer as end group, this constitutes an elegant way to functionalize the polymer. 3-Mercatopropionic acid confers a carboxylic acid end group to the PNIPAAm. By using 2-mercaptoethylamine as chain transfer agent under otherwise identical conditions, a PNIPAAm with an amino end group (PNIPAAm-NH$_2$) is produced. A biotin end group (PNIPAAm-biotin) can be produced by using 2-biotinamidoethanethiol (*see* **Subheading 3.1.**) as chain transfer agent.

2. NIPAAm from some suppliers contains an inhibitor to prevent accidental polymerization. This should be verified in each case, as only inhibitor-free NIPAAm can be used directly in the given protocol. The inhibitor can be removed from NIPAAm via recrystallization from hexane and vacuum drying at room temperature prior to use.

3. The indicated ^{1}H-NMR and mass data allow the positive identification of the synthesized molecule. The ^{1}H-NMR data were obtained with a WM 400 (400 MHz) FT spectrometer from Bruker Optics GmbH, Switzerland. The electrospray mass spectrometer used was an LCT from Micromass, England.

4. This procedure requires some experience and will seldomly succeed at first try. The most important aspect is the morphology of the precipitate. This should be a fine, easily filtratable solid. The morphology is adjusted via the amount of acetone used to dissolve the polymer. Too little acetone will result in a gel during precipitation that cannot be filtrated, too much acetone will not result in a precipitate at all (low yield). The amount of acetone indicated in **Subheading 3.2.** is therefore to be taken as an estimate, some fine tuning will have to take place in each case.

5. Compared to standard free radical polymerization, chain transfer polymerization should result in a narrow and evenly distributed molecular weight of the polymer preparation. This is important for affinity precipitation as the critical solution temperature depends on the molecular weight of the polymer. A broad distribution would therefore lead to fractionation during use. A polydispersity

$(P = M_w/M_n)$ below 1.5 as determined by MALDI-TOF MS is sufficiently narrow for a successful application of the polymer in affinity precipitation. Chain transfer polymerization also leads to smaller molecular weights (usually below 5000 g/mol) than standard free radical polymerization. This may have consequences for affinity precipitation, as precipitation becomes less quantitative with decreasing molecular weight. Preparations with an average molecular weight below 1000 g/mol should not be used. The average molecular weight of the PNIPAAm can be adjusted to some extent via the concentration of the chain transfer agent (less chain transfer agent leads to higher molecular weight). However, if the chain transfer agent concentration becomes too low, polymerization will no longer occur according to the chain transfer mechanism but rather by standard free radical polymerization. The indicated molar ratios between monomer, initiator, and chain transfer agent should therefore be used. Moreover, as the affinity ligand is inserted as an end group, the use of PNIPAAm oligomers assures a high ligand density per milligram of bioconjugate.

6. The molar ratio adjusted between the NHS-activated polymer and the avidin during coupling requires some fine tuning. A molar excess of NHS-activated polymer should be used in any case to assure a reasonable coupling yield, as PNIPAAm is known for its sluggish reactivity, and the need for an aqueous environment during coupling means that hydrolysis of the NHS groups constitutes an important side reaction. The exact ratio represents a balance between coupling yield in regard to the avidin and the number of remaining biotin-binding sites on the avidin (for subsequent linkage of the biotinylated affinity ligand). In our laboratory, we found, for example, a coupling yield of 41% together with an average number of free binding places of 3.4 per PNIPAAm-avidin when a ratio of 10:1 between PNIPAAm-NHS and avidin was used during coupling. This changed to a coupling yield of 71% (89%) and an average number of free binding places of 2.8 (1.9) when a ratio of 50:1 (100:1) was adjusted. A surplus of PNIPAAM-NHS during coupling means that some uncoupled PNIPAAm-COOH molecules remain in the preparation after hydrolysis of the unreacted NHS groups toward the end of the reaction. A quantification of this amount is possible via the avidin mass balance coupled with an end group titration (*see* **Subheading 3.9.**) after coupling. A removal of the unreacted PNIPAAm-COOH is also possible, for example, via affinity chromatography, but cumbersome. Since the unreacted PNIPAAm-COOH does not interfere with affinity precipitation, common practice is to simply leave these molecules in the final AML preparation. This has the additional advantage of increasing the absolute PNIPAAm concentration. For a quantitative precipitation, this should be 1% (w/w). If pure (and expensive) bioconjugates were used, this would correspond in most cases to a much too high capacity. In most large-scale applications, it is even common to add nonactivated PNIPAAm to the mixture in order to increase the total PNIPAAm concentration.

7. PNIPAAm has a critical solution temperature of 34°C in pure water. Efficient thermoprecipitation would therefore require 40°C, which will be too high in the

case of many biologicals. Salts lower the critical solution temperature in direct proportion to their 'salting out' capability. Ammonium sulphate is therefore a strong precipitation aiding agent. Adding the indicated amount (usually corresponding to 10% v/v of a saturated solution) will lower the critical temperature to approximately 20°C. Using a more biocompatible temperature of 30°C to achieve and maintain precipitation becomes then possible.

8. By using NHS biotin ester in this protocol, the synthesis of PNIPAAm-biotin becomes possible. However, the synthesis of PNIPAAm-biotin should preferably be done according to methods described in **Subheading 3.2.** using 2-biotinamidoethanethiol (*see* **Subheading 3.1.**) as chain transfer agent.

9. Vice versa it should in principle also be possible to couple a ligand consisting, for example, of a fusion protein bearing the biotin-binding domain of avidin to a PNIPAAm-biotin. However, to date, the use of such a molecule has not been described.

10. Affinity precipitation is relatively independent of the actual buffer composition. Often the conditions in the raw feed solution will be suitable to affinity complex formation. In such cases, the solid AML can be added directly to the feed in the desired concentration. Otherwise, buffer exchange must be performed, for example, via dialysis or ultrafiltration.

11. A 10-fold molar excess of binding places to target molecules usually works well. However, for fine tuning of the process, for example, in a large-scale bioseparation, it is usually worth the time to record an 'adsorption isotherm', that is, determine for a given amount of AML the distribution of bound and unbound target molecules as a function of the target molecule concentration over a wide concentration range from linear binding to complete saturation. This will give a very good idea of how much product will be recovered for a given amount of AML. In addition, the affinity constant can be determined from such a plot. Keep in mind that affinity precipitation is a one-stage process; there will always be an equilibrium between bound and unbound target molecules, although for a high-affinity AML this equilibrium may be well on the side of the affinity complex.

12. A 'binding temperature' during affinity complex formation that is 5°C below the critical temperature or a 'precipitation temperature' that is 5°C above this value usually work well (*see* **Subheading 3.10.** for determination of the critical solution temperature from the turbidity curve), as PNIPAAm shows a very sharp phase transition. However, in some solutions, phase transition may occur over a broader temperature interval. To be absolutely sure that a given temperature/solution will work, it is recommended to record the turbidity curve of the AML in that solution. Information concerning the critical solution temperature but also the broadness of the precipitation temperature interval may help to design a robust process. High salt concentrations have been described to broaden the phase transition temperature interval. In such cases, affinity precipitation is still possible, but the operating temperatures should be chosen to be farther away from the critical solution temperature. Also keep in mind that some PNIPAAm

bioconjugates show a phase transition hysteresis, that is, dissolution may occur at a lower temperature than precipitation.

13. A stirring of 30 min usually works well. However, the optimal duration of this process step is dictated by the kinetics of the affinity complex formation. For fine tuning of this parameter, the complex formation kinetics should be recorded, for example, by determining the concentration of the unbound target molecule in the supernatant as a function of time.

14. Centrifugation works well especially for small samples. However, the recovered gelatinous pellet is often difficult to dissolve. Repeated thermoprecipitation under binding conditions for 'washing', that is, the removal of physically entrapped impurities from the precipitate, may then become rather time consuming. In such cases, it may be helpful to add some fine cellulose fibres [e.g., Diacel 75 (75-μm fibres) from CFF GmbH + CoKG Gehren, Germany] in a concentration of 50% ($w/w_{polymer}$) to the solution. The precipitate will then redissolve quickly. The cellulose does not interfere with the affinity interaction. At larger scale, filtration may be more useful than centrifugation for precipitate recovery. The interested reader is referred to the polyTag Technology web page (http://www.polytag.ch) for protocols and instrumentation to be used in such cases.

15. Usually the pellet dissolves quicker at colder temperature. Especially for affinity complexes recovered by centrifugation in the absence of cellulose (*see* **Note 14** also), a dissolution temperature of 4°C is therefore recommended. For precipitates recovered in the presence of cellulose, a dissolution temperature of only a few degrees below the critical temperature (determination via turbidity curve in the dissolution buffer, keep the possibility of a hysteresis in mind and record the cooling down curve in this case) may be more suitable.

16. It is also possible to recover the affinity complex via filtration in the presence of cellulose (*see* **Note 14**). Then two methods are available for release of the target molecule. In the first case, cold dissolution buffer is passed through the filter cake thereby simultaneously redissolving the stimuli-responsive bioconjugate and releasing the target molecule from the complex. The other steps for target molecule recovery are described in **Subheading 3.6.** Alternatively, in this case also it is possible to pass a warm (temperature > critical temperature) dissociation buffer through the cake. In this case, the target molecule is released, while the stimuli-responsive AML stays in the cake.

17. Depending on the intended use, but also on the biochemistry of the interaction and the release conditions, the AML may be recycled for reuse. Validation of this possibility similar to procedures used for affinity chromatography stationary phases is recommended.

18. Avidin is a natural product; some variances in molecular weight and also in the extinction coefficient are unavoidable. The outlined procedure will give a good estimate. If a more precise determination is desired, the exact molecular weight should be determined by MS and the extinction coefficient by a calibration curve (absorbance at 280 nm) based on this molecular weight.

19. In the case of PNIPAAm-NH$_2$, the number of end groups can be determined by titration with 0.01 M perchloric acid/acetic acid using crystal violet as indicator. In the case of PNIPAAm biotin, the amount of biotin end groups can be determined by the HABA assay (*see* **Subheading 3.7.**).

20. MALDI-TOF MS is known to underestimate the broadness of the mass distribution. However, especially in the case of oligomeric PNIPAAm, this method gives quick and reliable results and is recommended.

21. Attention, most photometers with 'temperature control' indicate only the programmed temperature and do not measure the actual temperature in the measurement cell. However, for a correct estimation of the phase transition, knowledge of this temperature is absolutely necessary, so for the recording of the turbidity curves a photometer with in situ temperature measurement must be used.

22. The critical solution temperature of the PNIPAAm-based AML does not depend on the chemical nature of the affinity ligand or on the pH. Salts and many other cosolutes do have an influence. Salts normally lower the critical solution temperature. The effect is salt (cosolute) specific and may vary considerably from one type of salt to the next. This may be used for an intentional shifting of the LCST to the desired range (*see* **Note 7** also). The precipitation behaviour of the AML should always be characterized by measuring the turbidity curve in all reaction buffers prior to the design of the conditions for the actual affinity precipitation. Because the chemical nature of the molecules added as affinity ligand to the AML does not influence the precipitation behaviour, any AML precursor such as PNIPAAm-COOH or PNIPAAm-avidin can be used for the determination of the critical solution temperature and the general precipitation behaviour.

23. PNIPAAm shows rapid phase transition kinetics and a very narrow phase transition temperature interval. A heating rate of 0.5°C/min should therefore work. If at doubt, cloud point curves can also be recorded at lower heating rates, for example, 0.1°C/min. In the absence of kinetic effects, the two curves should be identical. If the turbidity decreases in the measured curves at elevated temperatures, the cause is normally not a redissolution of the precipitate but rather a sedimentation of the flocks.

24. For some PNIPAAms, for example, PNIPAAms with predominately isotactic structure, a pronounced hysteresis of the phase transition has been observed. In some cases, phase transition (redissolution) during cooling down occurred at a temperature that was 10°C below the phase transition (precipitation) temperature determined during heating. Because such a difference in the two critical temperatures may have consequences for the design of the affinity precipitation, the phase behaviour should be recorded for a full temperature cycle.

Acknowledgments

We would like to thank the numerous students and collaborators who helped to develop and test out the experimental procedures, in particular Matteo Costioli, Arnaud Deponds, Frédéric Garret-Flaudy, Marilia Panayiotou, and

Gisela Stocker. Our work in this area has for many years been supported by the German Science Foundation (DFG), the Swiss National Science Foundation (SNF), and the Commission for Innovation and Technology (CTI), Switzerland.

References

1. Green, N.M. (1975) Avidin. Adv. Protein. Chem. **29**, 85–113.
2. Hilbrig, F. and Freitag, R. (2003) Protein purification by affinity precipitation. J. Chromatogr. B. Analyt. Technol. Biomed. Life Sci. **790**, 79–90.
3. Schield, H.G. (1992) Poly(N-isopropylacrylamide) experiment, theory and application. Prog. Polym. Sci. **17**, 163–249.
4. Galaev, I.Y. and Mattiasson, B. (1993) Thermoreactive, water-soluble polymers, non-ionic surfactants, and hydrogels as reagents in biotechnology. Enzyme Microb. Technol. **15**, 354–366.
5. Freitag, R. and Garret-Flaudy, F. (2002) Salt effects on the thermoprecipitation of poly-(N-isopropylacrylamide)oligomers from aqueous solution. Langmuir **18**, 3434–3440.
6. Schield, H.G. and Tirrel, D.A (1991) Interaction of poly(N-isopropylacrylamide) with sodium n-alkyl sulfates in aqueous solution. Langmuir **7**, 665–671.
7. Galaev, I.Y. and Mattiasson, B. (1999) Smart polymers and what they could do in biotechnology and medicine. Trends Biotechnol. **17**, 335–340.
8. Garret-Flaudy, F. and Freitag, R. (2000) Use of the avidin-(imino)biotin system as a general approach to affinity precipitation. Biotechnol. Bioeng. **71**, 223–234.
9. Costioli, M.D., Fisch, I., Garret-Flaudy, F., Hilbrig, F. and Freitag R. (2003) DNA purification by triple-helix affinity precipitation. Biotechnol. Bioeng. **81**, 535–545.
10. Green, N.M. (1970) Spectrophotometric determination of avidin and biotin. Methods Enzymol. **18A**, 418–424.

5

Capturing Biotinylated Proteins and Peptides by Avidin Functional Affinity Electrophoresis

Bao-Shiang Lee, Sangeeth Krishnanchettiar, Syed Salman Lateef, and Shalini Gupta

Summary

Avidin functional affinity electrophoresis (AFAEP) is a variational method of affinity electrophoresis. In this technique, avidin is immobilized within a small area of the gel matrix by interaction with acrylamide and/or polyacrylamide either directly or through bifunctional linker glutaraldehyde during polymerization. Analytes can be heated with Tris–glycine sodium dodecyl sulfate (SDS) sample buffer so that biotinylated peptides/proteins are negatively charged and migrate electrophoretically towards the cathode through the avidin zone regardless of their isoelectric point (pI) values. Alternatively, if the behavior of the biotinylated analytes is known, the SDS treatment is not required. The polarity of the electrodes is set such that biotinylated analytes migrate electrophoretically through the avidin zone. This technique can work with or without SDS in gel running buffer. The AFAEP method allows the capture and concentration of biotinylated peptides/proteins. The values of this technique stem from a combination of merits of polyacrylamide gel electrophoresis and affinity technology.

Key Words: Avidin functional affinity electrophoresis; polyacrylamide gel electrophoresis; biotinylated proteins and peptides; isotope-coded affinity tags; glycoprotein; matrix-assisted laser desorption/ionization time of flight mass spectrometry.

1. Introduction

Avidin functional affinity electrophoresis (AFAEP) method *(1)* makes use of both biotin–avidin bonding and affinity electrophoresis (AEP) to capture the biotinylated peptides/proteins. Because the biotin–avidin complex bond formation is very rapid and is the strongest known non-covalent binding

From: *Methods in Molecular Biology, vol. 418: Avidin-Biotin Interactions, Methods and Applications*
Edited by: R. J. McMahon © Humana Press, Totowa, NJ

(dissociation constant of approximately 10^{-15} M), biotin–avidin interaction is widely used in a variety of biochemical techniques *(2)*. AEP *(3)* and its equivalent immunoelectrophoresis *(4)* have been used successfully for years in glycoprotein and antibody–antigen studies, respectively. In AFAEP, avidin is immobilized in a specific region of a polyacrylamide gel by interaction with acrylamide and/or polyacrylamide directly or with the help of a bifunctional linker glutaraldehyde during polymerization (*see* **Note 1**). Analytes are heated with Tris–glycine sodium dodecyl sulfate (SDS) sample buffer so that biotinylated peptides/proteins are negatively charged and migrate electrophoretically towards the cathode through the avidin zone regardless of their isoelectric point (pI) values. The SDS treatment and the presence of 0.1% (w/v) SDS in Tris–glycine electrophoresis running buffer do not influence the avidin–biotin bond formation *(5–7)*. As a result, the biotinylated peptides/proteins form a complex with avidin at the avidin zone, and no further migration occurs whereas non-biotinylated peptides/proteins migrate through the avidin zone freely. The AFAEP method allows one to capture and concentrate biotinylated peptides/proteins. Matrix-assisted laser desorption/ionization time of flight mass spectrometry (MALDI-TOF MS) is conveniently used to detect the avidin-captured biotinylated peptides/proteins. The AFAEP has many potential applications in biotin–avidin techniques. AFAEP is used successfully *(8)* in capturing the biotinylated tryptic peptides in the isotope-coded affinity tags technique *(9,10)* as an alternative to the avidin affnity column. Also, a new method of capturing glycoprotein by the AFAEP has been implemented (*see* **Note 2**). This method involves the use of biocytin hydrazide to couple to oxidized glycoproteins (periodate oxidation of carbohydrate's adjacent hydroxyls to generate highly reactive aldehydes) and subsequent capturing of the biotinylated glycoproteins with avidin.

2. Materials

2.1. Sample Preparation for AFAEP

1. Human biocytin-β-endorphin (cat. no. E 8139, Sigma, St. Louis, MO, USA; pI 9.6, m/z 3818, 1 mol biotin/mol endorphin), human -endorphin (cat. no. E 6261, Sigma; pI 9.6, m/z 3465), biocytin-protein kinase C fragment 19-31 (cat. no. P 8963, Sigma; pI 9.6, m/z 1914, 1 mol biotin/mol protein kinase C fragment 19-31), porcine biocytin-neuropeptide Y (cat. no. B 8530, Sigma; pI 8.3, m/z 4608, 1 mol biotin/mol neuropeptide Y), biotinylated bovine pancreas insulin (cat. no. I 2258, Sigma; pI 5.3, m/z 5735, 1-2 mol biotin/mol biotinylated insulin) are dissolved in SDS sample buffer at 5 µg/20 µl (*see* **Note 3**).
2. Biotinylated bovine serum albumin (BSA) (cat. no. A 8549, Sigma; 66 kDa, pI 5.5, 8–16 mol biotin/mol BSA) and non-biotinylated ovalbumin (cat. no.

A 2512, Sigma; chicken, 45 kDa, pI 5.1) are dissolved in SDS sample buffer at 15 µg/20 µl. (*see* **Note 3**).

3. Glycoprotein fetal calf serum futerin (cat. no. F 3004, Sigma), non-glycoprotein BSA (cat. no. A 7638, Sigma), and glycoprotein horseradish peroxidase (cat. no. 0031491, Pierce, Rockford, IL, USA) are first oxidized with sodium periodate and then incubated with biocytin hydrazide.

2.2. Avidin Functional Affinity Polyacrylamide Gel

1. Acrylamide/bis-acrylamide monomer stock solution [30.8% (w/v)] (cat. no. A 3699, Sigma): 37.5:1 Mix ratio, 30% T, 2.6% C. Stored at 4°C (*see* **Note 4**).
2. SDS [10% (w/v)] (cat. no. L 4522, Sigma).
3. Ammonium persulfate initiator solution: 10% (w/v) ammonium persulfate (cat. no. 431532, Sigma). Dissolve 0.1 g ammonium persulfate in an Eppendorf tube with deionized water (Millipore, Billerica, MA, USA) to 1.0 ml. Prepare fresh each time (*see* **Note 5**).
4. N,N,N′,N′-Tetramethylethylenediamine (TEMED; cat. no. 411019, Sigma).
5. Avidin sock solution: avidin (cat. no. A 9390, Sigma) is dissolved in deionized water at 10 µg/µl. Store in 20 µl aliquots at 4°C, used within a month (*see* **Note 6**).
6. Native resolving gel buffer (4×): 1.5 M Tris–HCl, pH 8.8. Store at room temperature.
7. SDS sample buffer (2×): 0.125 M Tris, 4% (w/v) SDS, 20% (v/v) glycerol, 0.02% (w/v) bromophenol blue, pH 6.8.
8. Native gel running buffer: 25 mM Tris, 192 mM glycine, pH 8.3. Store at room temperature.
9. SDS gel running buffer: 25 mM Tris, 192 mM glycine, 0.1% (w/v) SDS, pH 8.3. Store at room temperature.
10. Empty gel cassette (cat. no. NC2010, Invitrogen, Carlsbad, CA, USA] with dimension of 100 × 100 × 1 mm. (*see* **Note 7**).
11. Glutaraldehyde aqueous solution, 25% (cat. no. G 5882, Sigma), or 50% glutaraldehyde aqueous solution (cat. no. G 7651, Sigma).
12. Homemade plastic strips with dimension of 80 × 8 × 1 mm.
13. Centrifugal evaporator (SpeedVac concentrator) (Savant, Farmingdale, NY, USA).
14. PowerEase 500 programmable power supply and gel running box (Invitrogen).
15. Gel-staining solution: 0.1% Coomassie brilliant blue G250, 50% methanol, 10% acetic acid.
16. Destaining solution: 7% acetic acid, 12% methanol.

2.3. MALDI-TOF MS

1. Trifluroaetic acid (TFA), HPLC grade.
2. Acetonitrile (cat. no.271004, Sigma).

3. α-Cyano-4-hydroxycinnamic acid (CHCA) (cat. no. 203072, Brucker-Daltonics, Billerica, MA, USA) matrix solution: saturated solution (~10 mg/ml) in 50% aqueous acetonitrile containing 0.1% TFA. Prepare fresh each time.

4. Peptide extraction solution: 95% (v/v) formamide aqueous solution at pH 8.2 (*see* **Note 8**).

5. Trypsin (cat. no. V5111, Promega, Madison, WI, USA).

6. MALDI plate (cat. no. V700401, Applied Biosystems, Foster City, CA, USA; *see* **Note 9**).

7. Voyager-DE Pro mass spectrometer (Applied Biosystems) equipped with a 337-nm pulsed nitrogen laser.

8. ZipTips (cat. no. ZTC 18S960, Millipore) packed with C18 resin.

9. Vortex mixer (National Labnet, Woodbridge, NJ, USA).

10. Microcentrifuge (Jouan, Winchester, VA, USA).

3. Methods

AFAEP allows one to capture and concentrate biotinylated peptides/proteins heated with Tris–glycine SDS sample buffer with or without SDS in the gel running buffer. Presence of SDS does not influence avidin–biotin bonding. The 7.5% (w/v) polyacrylamide gel (*see* **Note 10**) is used in conducting AFAEP. Glutaraldehyde helps to incorporate the avidin into the gel matrix network by interaction with the amino/amide groups. The majority of avidin is confined to the incorporation area and remains immobile under the experimental conditions. Tris–glycine SDS sample buffer is used to ensure migration of the protein and peptide in one direction regardless of their pI values. The specific activity for biotin (MW 244) binding of avidin is approximately 14 μg/mg avidin, and the bond formation is very rapid. Therefore, 200 μg of avidin should be able to bind approximately 70 μg of biotinylated BSA *(1)*. MALDI-TOF MS was used to detect the molecular ions before and after AFAEP of human biocytin-β-endorphin (*see* **Fig. 1**), human β-endorphin (*see* **Fig. 1**), biocytin-protein kinase C fragment 19-31 *(11)*, porcine biocytin-neuropeptide Y *(11)*, and biotinylated bovine pancreas insulin *(11)*. Treating the biotinylated peptide–avidin gel piece with 95% (v/v) formamide aqueous solution (pH 8.2) for 20min at 65°C is sufficient to elute the biotinylated peptides. Biotinylated BSA stops its electrophoretic migration upon encountering avidin with or without SDS in the gel running buffer (*see* **Fig. 2**). Coomassie blue staining and peptide mass fingerprinting *(12)* by MALDI-TOF MS was used to confirm the capture of biotinylated BSA. However, electrophoretic mobility of non-biotinylated ovalbumin remains unaffected because it does not interact with avidin (*see* **Fig. 2**). These data confirm that biotinylated peptides/proteins stop its electrophoretic migration, whereas non-biotinylated peptides/proteins migrate through the avidin area. One interesting application of the AFAEP is

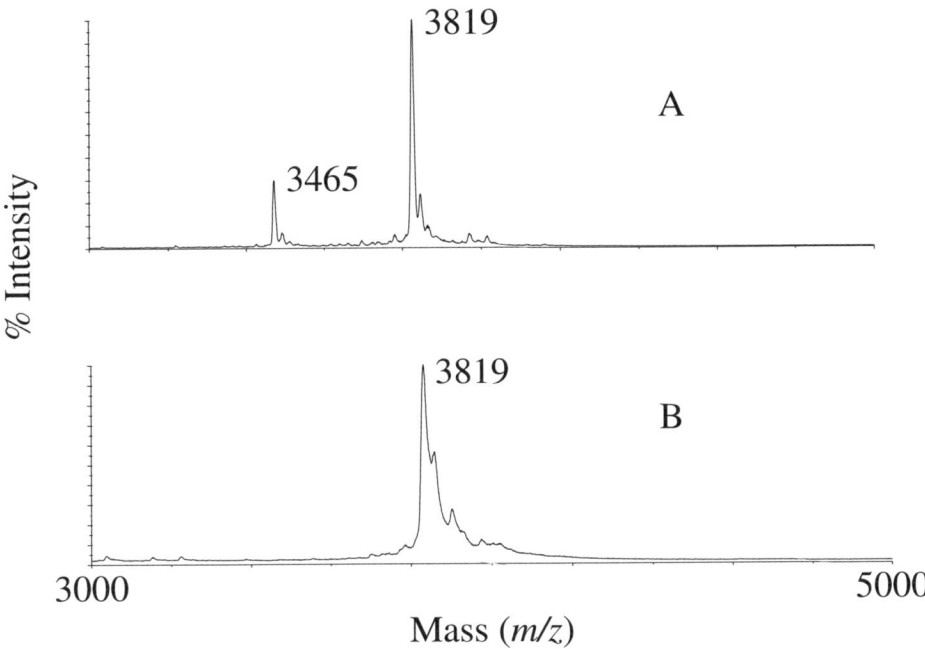

Fig. 1. Matrix-assisted laser desorption/ionization time of flight mass spectrum of the human biocytin-β-endorphin (5 μg) and non-biotinylated human β-endorphin (5 μg) before (**A**) and after (**B**) avidin functional affinity electrophoresis (AFAEP) with native gel running buffer *(1)*. Samples are heated at 95°C for 15 min with sodium dodecyl sulfate Tris–glycine sample buffer. The AFAEP-captured biotinylated peptides are excised from the gel and extracted with aqueous 95% formamide (pH 8.2) at 65°C for 20 min. ZipTips (Millipore), packed with C18 resin, are used to desalt biotinylated peptides prior to mass spectrometric analyses. Redrawn with permission from **ref.** *1*.

to capture glycoproteins (*see* **Fig. 3**). Here, carbohydrate moiety of glycoproteins futerin and horseradish peroxidase are been oxidized by periodate and subsequently biotinylated with biocytin hydrazide *(13)*. The biotinylated glycoprotein futerin and horseradish peroxidase are easily captured using AFAEP. However, the non-glycoprotein BSA is unaffected by this procedure.

3.1. Preparation and Running of Avidin Functional Affinity Electrophoresis

1. The native 7.5% (w/v) polyacrylamide resolving gel solution is made by mixing 7.5 ml of 30.8% (w/v) stock monomer acrylamide solution, 7.5 ml of 1.5 M Tris–HCl 4× resolving gel buffer (pH 8.8), 14.9 ml of deionized water, 150 μl of 10%

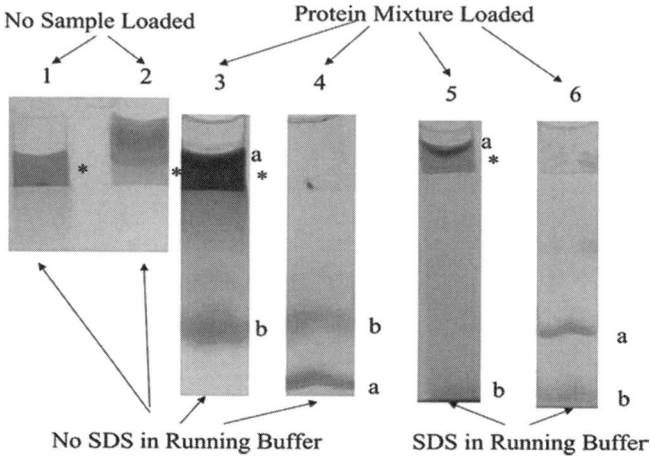

Fig. 2. Avidin functional affinity electrophoresis of biotinylated bovine serum albumin (BSA) and non-biotinylated ovalbumin (1). Mixtures of 15 µg biotinylated BSA (a) and non-biotinylated ovalbumin (b) are loaded onto lanes 3, 4, 5, and 6. Samples were heated at 95°C for 15 min with sodium dodecyl sulfate (SDS) Tris–glycine sample buffer. Native Tris–glycine gel running buffer is used in lanes 1, 2, 3, and 4. SDS Tris–glycine gel running buffer is used in lanes 5 and 6. Avidin (200 µg) is immobilized with bifunctional linker glutaraldehyde (lanes 1, 3, and 5) or without bifunctional linker glutaraldehyde (lane 2). Electrophoresis is done at approximately 125 V and 20 mA for roughly 1.25 h. Coomassie blue staining is used to visualize proteins. Asterisks indicate the position of the immobile avidin in the gel. Reproduced with permission from **ref. *1***.

   (w/v) ammonium persulfate solution, and 10 µl of tetramethylethylenediamine (TEMED). (*see* **Note 11**).
2. Native 7.5% (w/v) polyacrylamide resolving gel solution (8 ml) was poured into a 100 × 100 × 1-mm empty gel cassette, and sample wells of 22 × 8 × 1 mm are cast with plastic strips (80 × 8 × 1 mm).
3. Following gel formation, the plastic strips are removed.
4. The avidin-immobilized gel plug is formed by pouring 75 µl of the native 7.5% (w/v) polyacrylamide gel solution containing 2–5 µl of 25% (v/v) to 50% (v/v) glutaraldehyde aqueous solution and 200 µg of powder avidin into each sample well.
5. Gel plug without avidin (negative control) is formed by pouring 75 µl of the native 7.5% (w/v) polyacrylamide gel solution containing 2–5 µl of 25% (v/v) to 50% (v/v) glutaraldehyde aqueous solution.
6. Finally, 75 µl of native 7.5% (w/v) polyacrylamide gel is poured and polymerized to complete the sample wells.
7. Samples (5 µg each peptide and 15 µg each protein) are first heated at 95°C for 15 min in SDS Tris–glycine sample buffer.

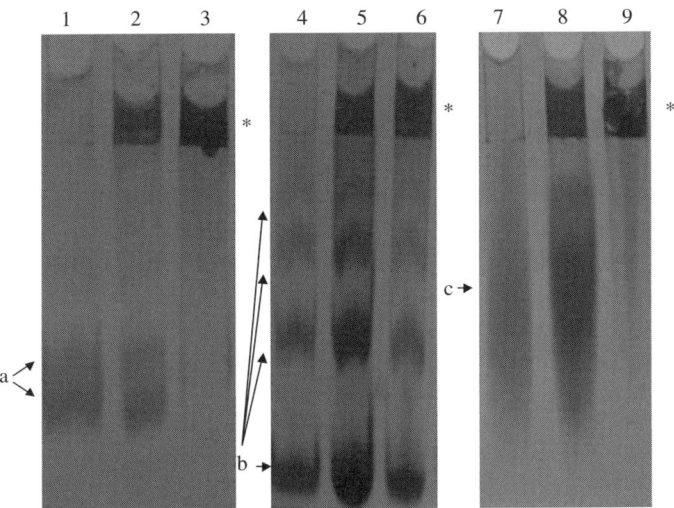

Fig. 3. Avidin functional affinity electrophoresis (AFAEP) of glycoprotein fetal calf serum futerin (a), non-glycoprotein bovine serum albumin (BSA) (b), and glycoprotein horseradish peroxidase (c) which have been incubated with biocytin hydrazide with or without prior oxidation with sodium periodate *(22)*. Lane 1, 150 µg futerin; lane 2, 150 µg futerin; lane 3, 150 µg futerin; lane 4, 150 µg BSA; lane 5, 300 µg BSA; lane 6, 150 µg BSA; lane 7, 150 µg horseradish peroxidase; lane 8, 300 µg horseradish peroxidase; lane 9, 150 µg horseradish peroxidase. Three hundred micrograms of avidin is immobilized with the bifunctional linker glutaraldehyde in lanes 2, 3, 5, 6, 8, and 9. Avidin is not embedded in lanes 1, 4, and 7 (negative controls). Protein samples for lanes 1, 3, 4, 6, 7, and 9 are oxidized with periodate and subsequently incubated with biocytin hydrzide. Protein samples for lanes 2, 5, and 8 are prepared without the periodate oxidation step. Samples are heated at 95°C for 15 min with SDS Tris–glycine sample buffer. A 7.5% native polyacrylamide gel and native Tris–glycine gel running buffer are used. Electrophoresis is done at 125 V and 20 mA for 1.5 h for lanes 1–6 and 2.5 h for lanes 7–9. Coomassie blue staining is used to visualize proteins. Asterisks indicate the position of the immobile avidin in the gel. The biotinylated glycoprotein futerin and horseradish peroxidase are easily captured by AFAEP. However, the non-glycoprotein BSA is unaffected by this procedure. Reproduced with permission from **ref**. *22*.

8. Native Tris–glycine gel running buffer, 500 ml, or SDS Tris–glycine gel running buffer is used for electrophoresis.
9. PowerEase 500 programmable power supply (Invitrogen) is used to run the gel.
10. Electrophoresis is carried out at 3 W with a current of 22 mA and a voltage of 125 V for 1.25–2.5 h.
11. Gels are stained with Coomassie blue G-250 and documented using a digital camera and a light box.

3.2. Detection of AFAEP-Captured Biotinylated Peptides/Proteins by Mass Spectrometry

1. The human biocytin-β-endorphin, non-biotinylated human β-endorphin, biocytin-protein kinase C fragment 19-31, porcine biocytin-neuropeptide Y, biotinylated bovine pancreas insulin, biotinylated BSA, and non-biotinylated ovalbumin (5 µg each peptide and 15 µg each protein) are used as analytes.
2. The gel plug containing the avidin-biotinylated peptides is excised and the biotinylated peptides are extracted using 200 µl peptide extraction solution at 65°C for 20 min.
3. ZipTips, packed with C18 resin, are used to prepare the extracted peptide solution for mass spectrometry analysis using CHCA as the matrix.
4. Aliquots (1.3 µl) of the matrix solution is used to elute the peptide from ZipTips and spotted onto a MALDI-TOF target and analyzed by a positive ion Voyager-DE PRO mass spectrometer equipped with a nitrogen laser.
5. Peptide mass is measured to confirm the captured biotinylated peptides using a linear mode and a external mass calibration using the peaks of a mixture of bradykinin fragments 1–7 at m/z 758, angiotensin II (human) at m/z 1047, $P_{14}R$ (synthetic peptide) at m/z 1534, adrenocorticotropic hormone fragment18-39 (human) at m/z 2467, insulin oxidized B (bovine) at m/z 3497, insulin (bovine) at m/z 5735.
6. The gel plug containing the avidin-biotinylated proteins is excised and digested by trypsin using a standard protocol *(12)*. MALDI-TOF mass fingerprinting or LC-MS/MS is used to confirm the captured biotinylated proteins.

3.3. Capturing Glycoproteins by AFAEP

1. Dissolve 1 mg horseradish peroxidase, futerin, or BSA in 100 µl 20 mM ammonium acetate aqueous solution containing 10 mM sodium periodate, pH 5.5.
2. Incubate the sample for 30 min at room temperature in the dark.
3. Apply the protein solution to a Micro Bio-Spin 6 desalting column (Bio-Rad, Hercules, CA, USA) using 20 mM ammonium acetate aqueous solution, pH 5.5.
4. Add biocytin hydrazide (cat. no. 280200, Pierce) to a final concentration of 10 mM and incubate for 1 h at room temperature.
5. Apply the protein solution to a Micro Bio-Spin 6 desalting column (Bio-Rad) using deionized water and then dry the solution in a centrifugal evaporator.
6. Sample is heated at 95°C for 15 min in SDS Tris–glycine sample buffer.
7. AFAEP is used to capture the glycoproteins (*see* **Fig. 3**).
8. Coomassie blue staining is used to visualize the proteins.

4. Notes

1. It has been observed in our work *(14–16)* and those of others *(17–19)* that adding proteins to the acrylamide gel solution directly and thereby allowing the gel to be polymerized, some protein gets covalently and/or non-covalently incorporated into the gel matrix network possibly through interactions with the double bonds

of acrylamide during polymerization. The bifunctional linker glutaraldehyde can also be added to incorporate the protein into gel matrix network by interaction with the amino/amide groups of acrylamide and/or polyacrylamide *(1,17,20,21)* during polymerization. It is also possible that glutaraldehyde causes crosslinking of some of the protein molecules, and the crosslinked protein molecules get trapped in the gel pore.

2. It is interesting to note that while developing a new method of capturing glyco-protein *(22)*, we discover that hydrazide, which has been used extensively in capturing glycoprotein, can react with non-glycosylated proteins/peptides at certain conditions and may create false-positive results *(23)*. Extra precaution should be exercised to reduce this side reaction.

3. Pure solid peptides/proteins can be dissolved in SDS sample buffer directly at 1–0.1 mg/ml. Do not use protein solutions that are at extreme pH. Should the sample solution turn from blue to yellow after adding sample buffer, the sample is too acidic.

4. Acrylamide is a neurotoxin when unpolymerized and should be handled with great care.

5. It is critical to use deionized water (conductivity >18 MΩ) for all solutions to minimize interference by the impurities present in water.

6. Each aliquot of avidin (tetrameric glycoprotein) stock solution is dried immediately before using. Strepavidin (cat. no. 21125, Pierce) and neutravidin (cat. no. 31000, Pierce) can be used as alternatives to avidin (more economical). Strepavidin and neutravidin are carbohydrate free with a near neutral pI with less solubility in water than avidin.

7. Eight millilitres of polyacrylamide gel solution is enough for one 1-mm-thick empty gel cassette. Empty gel cassette (cat. no. NC2015, Invitrogen) with dimension of 100 × 100 × 1.5 mm can be used for large sample loading.

8. Other peptide extraction solution such as aqueous 6 M guanidine HCl (pH 1.5) or even water can also be used *(8,11)*.

9. Hundred-well gold-coated stainless flat plate is good for routine multiple sample analysis with close external calibration using peptides/proteins of know m/z values on an adjacent spot. Teflon-coated plate, 384-well, is particularly suited for concentrating samples for increased sensitivity due to its small spot size.

10. Different percent gel from 5 to 15% or gradient gel with or without stacking gel can be used.

11. Degassing prevents oxygen in solution from reacting with free radicals and inhibiting polymerization. As heat liberated from the polymerization process can result in the formation of small air bubbles in the gel, it is advisable to perform degassing as a matter of routine.

Acknowledgment

This work was supported by Research Resources Center at the University of Illinois at Chicago.

References

1. Lee, B., Gupta, S., Krishnanchettiar, S., and Lateef, S. S. (2005) Capturing SDS treated biotinylated protein and peptide by avidin functional affinity electrophoresis with or without SDS in the gel running buffer. *Anal. Biochem.* **336,** 312–315.
2. Wilchek, M. and Bayer, E. A., eds. (1990) *Avidin–Biotin Technology.* Academic Press, San Diego.
3. Bog-Hansen, T. C. and Hau, J. (1982) Affinity electrophoresis of glycoproteins. *Acta Histochem.* **71,** 47–56.
4. Nakamura, K. and Takeo, K. (1998) Affinity electrophoresis and its applications to studies of immune response. *J. Chromatogr.* **715,** 125–136.
5. Karlin, A., Wang, C., Li, J., and Xu, Q. (2004) Transfer in SDS of biotinylated proteins from acrylamide gels to an avidin-coated membrane filter. *Biotechniques* 36, 1010–1016.
6. Bayer, E., Ehrlich-Rogozinski, S., and Wilchek, M. (1996) Sodium dodecyl sulfate polyacrylamide gel electrophoresis method for assessing the quaternary state and comparative thermostability of avidin and streptavidin. *Electrophoresis* **17,** 1319–1324.
7. Lee, H. and Fritsche, E. (2003) Determination of biotin on a protein by quanti-tative sodium dodecyl sulfate capillary gel electrophoresis of monomeric avidin. *J. Chromatogr. A* **994,** 213–219.
8. Lee, B., Krishnanchettiar, S., Lateef, S. S., and Gupta, S. (2006) Isotope-coded affinity technique using avidin functional affinity electrophoresis: an alternative to an avidin affinity column. *J. Chin. Chem. Soc.* **53,** 745–750.
9. Gigy, S., Rist, B., Gerber, S., Turecek, F., Gelb, M., and Aebersold, R. (1999) Quantitative analysis of complex protein mixtures using isotope coded affinity tags. *Nat. Biotechnol.* **17,** 994–999.
10. Aebersold, R. and Mann, M. (2003) Mass spectrometry-based proteomics. *Nature* **422,** 198–207.
11. Lee, B., Krishnanchettiar, S., Lateef, S. S., and Gupta, S. (2005) Mass spectrometric detection of biotinylated peptides captured by avidin functional affinity. *Rapid Commun. Mass Spectrom.* **19,** 886–892.
12. Kinter, M. and Sherman, N. E. (2000) *Protein Sequencing and Identification Using Tandem Mass Spectrometry.* Wiley & Sons, New York.
13. Hermanson, G. T. (1996) *Bioconjugate Techniques.* Academic Press, San Diego.
14. Lee, B., Gupta, S., Krishnanchettiar, S., and Lateef, S. S. (2004) Countercurrent affinity electrophoresis of biotinylated proteins. *Anal. Biochem.* **330,** 178–180.
15. Lee, B., Gupta, S., Krishnanchettiar, S., and Lateef, S. S. (2004) Catching protein antigens by antibody affinity electrophoresis. *Electrophoresis* **25,** 3331–3335.
16. Lee, B., Gupta, S., Krishnanchettiar, S., and Lateef, S. S. (2004) Catching and separating protein ligands by functional affinity electrophoresis. *Anal. Biochem.* **334,** 106–110.
17. Somani, B., Ambade, V., and Arora, M. (2003) Polyacrylamide gel affinity electrophoresis for separation of enzyme isoforms. *Med. J. Armed Forces India* **59,** 125–127.

18. Bonaventura, C., Bonaventura, J., Stevens, R., and Millington, D. (1994) Acrylamide in polyacrylamide gels can modified proteins during electrophoresis. *Anal. Biochem.* **222,** 44–48.

19. Chiari, M., Righetti, P., Negri, A., Ceciliani, F., and Ronchi, S. (1992) Preincubation with cystein prevents modification of SH groups in proteins by unreacted acrylamide in a gel. *Electrophoresis* **13,** 882.

20. Lee, B., Krishnanchettiar, S., Lateef, S. S., and Gupta, S. (2005) Capturing sodium dodecyl sulfate-treated protein antigens by antibody affinity electrophoresis. *Electrophoresis* **26,** 501–513.

21. Patrica, J. M., Lisa, R. B., Lucinda, J. W., Luz, M. N., and Ian, H. M. (1990) Affinity purification of polyclonal antibodies from antigen immobilized in situ in SDS polyacrylamide gels. *Anal. Biochem.* **187,** 244–250.

22. Lee, B., Krishnanchettiar, S., Lateef, S. S., and Gupta, S. (2006) New methods of affinity electrophoresis. *Curr. Anal. Chem.* **2,** 243–251.

23. Lee, B., Krishnanchettiar, S., Lateef, S. S., and Gupta, S. (2007) Biotinylation of peptides/proteins using biocytin hydrazide. *J. Chin. Chem. Soc.* **54,** 541–548.

6

Functionality Screen of Streptavidin Mutants by Non-Denaturing SDS–PAGE Using Biotin-4-Fluorescein

Nicolas Humbert and Thomas R. Ward

Summary

Site-directed mutagenesis or directed evolution of proteins often leads to the production of inactive mutants. For streptavidin and related proteins, mutations may lead to the loss of their biotin-binding properties. With high-throughput screening methodologies in mind, it is imperative to detect, prior to the high-density protein production, the bacteria that produce non-functional streptavidin isoforms. Based on the incorporation of biotin-4-fluorescein in streptavidin mutants present in *Escherichia coli* bacterial extracts, we detail a functional screen that allows the identification of biotin-binding streptavidin variants. Bacteria are cultivated in a small volume, followed by a rapid treatment of the cells; biotin-4-fluorescein is added to the bacterial extract and loaded on an Sodium Dodecyl Sulfate Poly-Acrylamide Gel Electrophoresis (SDS–PAGE) under non-denaturing conditions. Revealing is performed using a UV transilluminator. This screen is thus easy to implement, cheap and requires only readily available equipment.

Key Words: Streptavidin; biotin-4-fluorescein; functionality screening; *E. coli*; fluorescence quenching; SDS–PAGE.

1. Introduction

For the past 30 years, the biotin–(strept)avidin system has found widespread applications in various biotechnology-related fields. Among these, one should mention immunolabeling, affinity targeting, analytical biochemistry, drug targeting and chemistry *(1–11)*. For certain applications (i.e., medical applications), it would be particularly useful to generate (strept)avidin mutants

From: *Methods in Molecular Biology, vol. 418: Avidin-Biotin Interactions, Methods and Applications*
Edited by: R. J. McMahon © Humana Press, Totowa, NJ

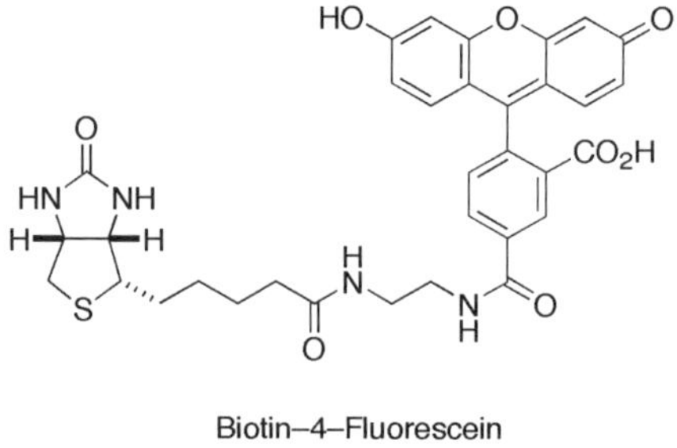

Biotin–4–Fluorescein

Scheme 1. Structure of the biotin-4-fluorescein.

[(strept)avidin refers to either avidin or streptavidin] which display enhanced affinity for biotinylated probes compared to unfunctionalized biotin *(2,3,9–11)*.

For this purpose, mutants have been produced, either by site-directed or by directed-evolution techniques, and it would be desirable to rapidly screen for the biotin-binding ability of the produced (strept)avidin isoforms. In our experience, the random generation of mutants yields a large number of misfolded and/or inactive mutants *(12)*. The identification of the inactive isoforms should be possible by the development of a functionality screen combined with enzymatic assays *(13,14)*. This kind of screening should be rapid, straightforward, cheap, reliable and highly sensitive.

Fluorescence appears as a versatile and highly sensitive means to detect molecules, as it can be revealed by UV analysis. As demonstrated by Gruber, the biotin-4-fluorescein is a convenient probe for streptavidin assays (*see* **Scheme 1**). This "fluorescent biotin" allows to quantify the amount of (strept)avidin present in a crude biofluids and allows the determination of the number of active sites of pure (strept)avidin *(15)*. We have recently shown that biotin-4-fluorescein can be used to detect the functionality of streptavidin in a bacterial crude extract applied on non-denaturing SDS–PAGE and revealed by a UV transilluminator *(16)*. Herein, we detail this technique.

2. Materials

2.1. Stock Solutions of Pure Streptavidin and Biotin-4-Fluorescein

1. Streptavidin stock solution (152 µM = 10 mg/mL, 65 700 g/mol). The pure protein (10.0 mg) is weighed in a 1.5-mL microcentrifuge tube and dissolved in nanopure water (1 mL). The solution is aliquoted (50 µL in 500-µL microcentrifuge tubes) and stored for up to 1 year at –20°C. The exact concentration can be determined by UV-Vis absorption but is not required (*see* **Notes 1** and **2**).
2. Biotin-4-fluorescein stock solution (644.7 g/mol): The biotin-4-fluorescein (Molecular Probes) is dissolved in dimethyl sulfoxide (final concentration =

0.6 mM). It is aliquoted in 500-μL microcentrifuge tubes (fractions of 100 μL) and stored at –80°C *(15)*.

2.2. Cell Culture and Lysis

1. Stock solution of glucose (20%, 180.2 g/mol): 200 g of D-glucose is dissolved in distilled water (final volume = 1 L), the solution is divided into 10 aliquots of 100 mL in autoclavable bottles, then autoclaved (20 min., 120°C, 1.5 bar). This solution is stored at room temperature for up to a few months.
2. Ampicillin stock solution (50 mg/mL): Ampicillin sodium salt (500 mg) is diluted in nanopure water to a final volume of 10 mL. The solution is aliquoted in 1 mL fractions and stored at –20°C for up to a few months *(17)*.
3. Chloramphenicol stock solution (34 mg/mL): Chloramphenicol (340 mg) is diluted in 70% ethanol solution to a final volume of 10 mL. The solution is aliquoted in 1 mL fractions and stored at –20°C for up to a few months *(17)*.
4. Isopropyl β-D-Thiogalactopyranoside (IPTG) stock solution: IPTG powder (MW = 238.3 g/mol) is weighed and dissolved in nanopure water to a final concentration of 0.8 M. The solution is then aliquoted in 1 mL fractions and stored at –80°C for up to a few months *(17)*.
5. IPTG working solution: The stock solution is diluted 10 times in nanopure water (final concentration = 80 mM). It can be stored at –20°C for a few weeks, but it is better to prepare it freshly.
6. Luria Broth medium supplemented with ampicillin, chloramphenicol and glucose (LACG) Petri dishes: The LB powder (Brunschwig) is mixed together with bacto-agar (Difco) and is dissolved with approximately 450 mL of distilled water in a 500-mL autoclavable bottle. After autoclaving (*see* **Note 3**, 20 min, 120°C, 1.5 bar), it is allowed to cool to 50–60°C. The stock glucose (25 mL), the ampicillin (600 μL of stock solution) and the chloramphenicol solution (500 μL) are added (final concentrations: 1%, 60 μg/mL and 34 μg/mL, respectively). The volume is completed to 500 mL with sterile nanopure water. The medium is mixed and 10–15 mL is poured out on each Petri dish. When the media are solidified, dishes are stored at 4°C for up to 3–4 months (*see* **Note 4**).
7. Modified TP (MTP) medium: Bactotryptone (10 g), Na_2HPO_4 (650 mg), KH_2PO_4 (500 mg), NaCl (4 g) and bacto-yeast extract are mixed together, dissolved in distilled water (500 mL), autoclaved (20 min, 120°C, 1.5 bar) and stored at room temperature for up to 1 year.
8. Preculture medium: The MTP medium (50 mL) is completed with 60 μL of ampicillin stock solution (final concentration = 60 μg/mL), 50 μL of chloramphenicol solution (final concentration = 34 μg/mL) and 2.5 mL of 20% glucose (final concentration = 1%). This medium can be stored for up to 1 week at 4°C.
9. Culture medium: The MTP (50 mL) is completed with 60 μL of ampicillin stock solution (final concentration = 60 μg/mL), 50 μL of chloramphenicol solution (final concentration = 34 μg/mL) and 1 mL of 20% glucose (final concentration = 0.4%). This medium can be stored for up to 1 week at 4°C.

2.3. Treatment of the Bacterial Extracts

1. Resuspending buffer: To a solution of 20 mM Tris–HCl pH 7.4, 0.02 w/v sodium azide and 10 mM, $MgCl_2$, add lyophilized DNase I (2–3 mg per 50 mL of solution). The solution can be stored at 4°C for up to 3 weeks.

2.4. SDS–PAGE

1. Separating buffer (4×): 1.5 M Tris–HCl, pH 8.8. Store at 4°C.
2. Stacking buffer (4×): 0.5 M Tris–HCl, pH 6.8. Store at 4°C.
3. Acrylamide/bisacrylamide solution (37.5:1) was purchased from National Diagnostics. It should be manipulated with a great care due to its high toxicity (neurotoxic and carcinogenic). It should be stored in a well-ventilated hood at room temperature.
4. SDS solution, 10%: As SDS is highly volatile and irritant, it is strongly advised to weigh it wearing glasses, gloves and a mask. Dissolve it in distilled water and store at room temperature.
5. N,N,N′,N′-Tetramethyl-ethylenediamide (TEMED). Store at 4°C.
6. Ammonium persulfate: Prepare a 10% (w/v) solution in distilled water, store at 4°C for immediate use or at –20°C for long-term storage (*see* **Note 5**).
7. Non-denaturing gel loading buffer (5×): 0.25 M Tris–HCl, pH 6.8, 0.03% (w/v) bromophenol Blue, 50% sucrose. Store at 4°C.
8. Running gel buffer (10×): 0.25 M Tris, 1.92 M glycine, 1% (w/v) SDS. Store at room temperature.

3. Methods

As biotin-4-fluorescein tends to diffuse out of the gel, it is necessary to perform the analysis by UV transilluminator immediately at the end of the migration of the gel. Due to the low volumes of bacterial precultures and cultures, a strict sterile environment is required at least until the induction of the cells. As the technique described here can be performed on a large amount of mutants, it is advised to use a multi-stepper pipette to perform the redundant steps (distribution of the medium in falcon tubes, induction, resuspension of the cell extracts).

3.1. Cell Cultures

1. Freshly transformed BL21(DE3)pLysS cells with either a wild-type, a mutated or by an empty plasmid (pET11b, Novagen) are plated on LACG dishes and incubated overnight (12–16 h) at 37°C. The dishes are sealed with a parafilm strip and stored at 4°C for up to 1 week (*see* **Note 6**).
2. One single circular colony (*see* **Note 7**) is picked and used to inoculate the preculture medium (1 mL) in 12-mL falcon tubes. The tubes are incubated overnight (12–16 h) at 37°C, under orbital shaking (250 rpm). To avoid contamination by bacteria having lost their plasmid, the preculture should be used immediately after incubation.

3. The resulting preculture (80 μL) is used to inoculate the culture medium (4 mL) in a 50-mL falcon tube (*see* **Note 8**). The culture is incubated 3 h at 37°C under orbital shaking (250 rpm).
4. Each miniculture is induced with IPTG (20 μL of the working solution).
5. The minicultures are subsequently incubated at 37°C under orbital shaking (250 rpm) for 2 h.
6. Centrifugation (5 min, 5000 g, room temperature) is performed to isolate the pellet which is stored at –20°C until use.

3.2. Treatment of the Samples Before SDS–PAGE

1. The cell pellets are thawed and the resuspension buffer (600 μL) is added. The bacterial extract is resuspended by gentle stirring.
2. Having totally digested the nucleic acids, the samples are transferred in a 1.5-mL Eppendorf tube and centrifuged (5 min at 16 000 g at room temperature).
3. The supernatant is transferred in a new 1.5-mL Eppendorf tube and the cell debris pellet is discarded.
4. Each extract (12 μL) is transferred in a 500-μL microcentrifuge tube, and the biotin-4-fluorescein stock solution (1 μL) as well as 5× non-denaturing gel loading buffer (3 μL) are added (*see* **Note 9**).
5. As positive control, the streptavidin solution stock (1.2 μL) is transferred in a 500-μL microcentrifuge tube. Nanopure water (10.8 μL), biotin-4-fluorescein stock solution (1 μL) and 5× non-denaturing gel loading buffer (3 μL) are added.

3.3. SDS–PAGE

1. These instructions assume the use of miniprotean 3 gel system (Bio-Rad), but are valid with other systems that allow the loading of 15 μL samples. Clean two to three times the glasses with a wet tissue (distilled water) to remove polyacrylamide residues of previous experiments, rinse with ethanol and air-dry.
2. Prepare a 1.5-mm-thick 16% running gel by mixing the 4× separating buffer (2.5 mL) with the acrylamide/bisacrylamide solution (5.3 mL), 10% SDS solution (100 μL) in 2.05 mL distilled water. Following this, add 10% ammonium persulfate (34 μL) and TEMED (5 μL) and mix the solution before pouring the gel immediately up to 2–2.5 cm under the higher border, leaving space for the stacking gel *(18)*. Overlay the gel mixture with distilled water or isopropanol. The polymerization of the gel should be complete within 40 min at room temperature.
3. Remove the supernatant liquid and rinse the top of the gel twice with distilled water.
4. Prepare the stacking gel by mixing the 4× stacking buffer (1.5 mL) with the acrylamide/bisacrylamide solution (1 mL), the 10% SDS (60 μL) and water (3.4 mL). As above, the 10% ammonium persulfate (60 μL) and the TEMED (9 μL) should be added last. Mix the solution, pour the gel immediately and place the comb gently.

5. Prepare the 1× gel running buffer (1 L) by diluting 10 times the 10× stock solution in a 1-L bottle. Mix by inverting the bottle two to three times.

6. Once the stacking gel is polymerized, remove carefully the comb and rinse twice the wells with 1× gel running buffer using a 1-mL pipette.

7. After assembling of the miniprotean gel, ensure the water tightness, place the gel in the chamber and fill the chamber with the 1× gel running buffer.

8. Load the samples (including a protein marker) with a micropipette.

9. Complete the assembly and connect the gel unit to a power supply. Run the gel at constant voltage (200 V) and stop when the loading buffer dye is on the bottom of the gel (45–60 min).

10. Disassemble the setup and put the gel in a small flat container containing distilled water.

3.4. UV Transilluminator Revealing

1. Put the gel directly on the transilluminator bottom. As a transilluminator is often used to reveal ethidium bromide agarose gels, it is strongly recommended to manipulate the SDS–PAGE with gloves.

2. Focus the camera on the gel under the daylight illumination and switch the UV light on (no filter required). As the fluorescence of the biotin-4-fluorescein is partially quenched by the protein, prolonged UV exposure is required (between

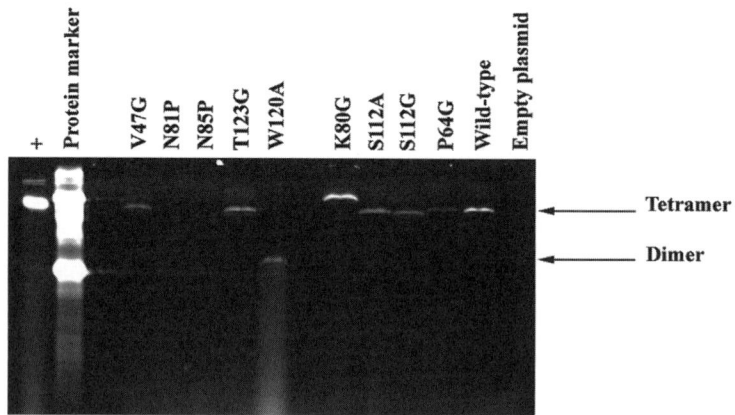

Fig. 1. Functionality screen of 10 streptavidin isoforms. One single band appears near 66 kDa (checked by Coomassie Blue staining, not shown), with the purified wild-type streptavidin (+). A similar band appears for all active mutants. No band is observed for either the empty plasmid showing that no bacterial compound interact with the biotin-4-fluorescein, or N81P and N85P mutants, suggesting that these two mutants are non-functional. K80G migrates less than the other mutants suggesting a strong modification of its quaternary structure. W120A migrates at approximately 33 kDa, confirming that this mutant is dimeric *(20)*.

0.5 s and 8 s depending on the power of the bulb). **Figure 1** shows an example of gel obtained with biotin-4-fluorescein applied on selected streptavidin mutant cell extracts.

4. Notes

1. As streptavidin has a low sensitivity to proteases, it can be stored in nanopure water and does not require buffered or saline solutions *(19)*. Nevertheless, many protocols refer to a storage of the protein in a Tris–HCl buffered solution (10 mM, pH = 7.4).

2. To determine the streptavidin concentration by UV-Vis absorption, transfer 10 μL of streptavidin stock solution in a new 1.5-mL microcentrifuge tube and add 740 μL of nanopure water (final protein concentration ~2 μM = the concentration used to quantify the active sites of the protein by fluorimetric titration) *(15)*. Measure the absorbance in a UV/V is spectrophotometer using a 1 cm quartz cuvette at a wave length of 280 nm ($\varepsilon_{SAV}^{280nm} = 167, 280/cm/M$)

3. After autoclaving, if antibiotics and glucose have not been added, the medium can be stored as a solid at room temperature in the bottle; it can be melted by microwave heating and then treated as described in **Subheading 2**. Another alternative consists to store the medium at 55–60°C where the agar is not solidified; in this case, glucose and antibiotic can be directly added and the medium can be poured in the Petri dishes.

4. To ensure the sterility of the Petri dishes, incubate one of them at 37°C overnight. If one or more bacterial colonies appear, the batch of dishes is probably contaminated and should be discarded.

5. Ammonium persulfate solution is usually stored at –20°C, but if it is frequently used, it can be stored at 4°C for 2–3 weeks. The solution should be discarded when the polymerization of the gcl takes longer than 1 h.

6. It is not obligatory to use a fresh transformation for wild-type and empty plasmid, and bacteria coming from a glycerol stock can be plated on the Petri dish. Moreover, if the freshly transformed bacteria are not used during the week following their transformation, a single colony can be picked to plate a new dish.

7. Ideally, the picked colony should be as circular as possible (which is not always the case because BL21(DE3)pLysS cells tend to form irregular colonies). Concerning the colony size, it should be 1–3 mm in diameter. If they are too big, a large amount of plasmid-lost bacteria may be present; if they are too small, they may be satellite colonies.

8. The cultures can be performed in a 24-well microplate for 2-mL cultures. In this case, all the volumes should be divided by two.

9. These two steps should be performed immediately prior to loading the samples on the SDS–PAGE, that is, the gel should be ready for loading "before" mixing the biotin-4-fluorescein with the bacterial extracts.

Acknowledgments

We thank Prof. C. R. Cantor for the streptavidin gene as well as Profs. P. Schürmann and J.-M. Neuhaus for their help in setting up the protein production. This work was funded by the Swiss National Science Foundation (Grants FN 620-57866.99 and FN 200021-105192/1 as well as NRP 47 "Supramolecular Functional Materials"), CERC3 (Grant FN20C321-101071), the Roche Foundation, the Canton of Neuchâtel as well as FP6 Marie Curie Research Training Network (IBAAC network, MRTN-CT-2003-505020).

References

1. Wilchek, M. & Bayer, E. A. (1999) Foreword and introduction to the book (strept)avidin-biotin system, *Biomol. Eng.* **16**, 1–4.
2. Hamblett, K. J., Kegley, B. B., Hamlin, D. K., Chyan, M. K., Hyre, D. E., Press, O. W., Wilbur, D. S. & Stayton, P. S. (2002) A streptavidin-biotin binding system that minimizes blocking by endogenous biotin, *Bioconjug. Chem.* **13**, 588–98.
3. Wilbur, D. S., Pathare, P. M., Hamlin, D. K., Stayton, P. S., To, R., Klumb, L. A., Buhler, K. R. & Vessella, R. L. (1999) Development of new biotin/streptavidin reagents for pretargeting, *Biomol. Eng.* **16**, 113–8.
4. van Osdol, W. W., Sung, C., Dedrick, R. L. & Weinstein, J. N. (1993) A distributed pharmacokinetic model of two-step imaging and treatment protocols: application to streptavidin-conjugated monoclonal antibodies and radiolabeled biotin, *J. Nucl. Med.* **34**, 1552–64.
5. Nakamura, M., Tsumoto, K., Ishimura, K. & Kumagai, I. (2002) Detection of biotinylated proteins in polyacrylamide gels using an avidin-fluorescein conjugate, *Anal. Biochem.* **304**, 231–5.
6. Lee, B. S., Gupta, S., Krishnanchettiar, S. & Lateef, S. S. (2005) Capturing SDS-treated biotinylated protein and peptide by avidin functional affinity electrophoresis with or without SDS in the gel running buffer, *Anal. Biochem.* **336**, 312–5.
7. Fukushima, H., Morgan, H. & Taylor, D. M. (1994) Self-assembly of avidin and streptavidin with multifunctional biotin molecules, *Thin Solid Films* **244**, 789–93.
8. Diamandis, E. P. & Christopoulos, T. K. (1991) The biotin-(strept)avidin system: principles and applications in biotechnology, *Clin. Chem.* **37**, 625–36.
9. Collot, J., Gradinaru, J., Humbert, N., Skander, M., Zocchi, A. & Ward, T. R. (2003) Artificial metalloenzymes for enantioselective catalysis based on biotin-avidin, *J. Am. Chem. Soc.* **125**, 9030–1.
10. Skander, M., Humbert, N., Collot, J., Gradinaru, J., Klein, G., Loosli, A., Sauser, J., Zocchi, A., Gilardoni, F. & Ward, T. R. (2004) Artificial metalloenzymes: (strept)avidin as host for enantioselective hydrogenation by achiral biotinylated rhodium-diphosphine complexes, *J. Am. Chem. Soc.* **126**, 14411–8.
11. Letondor, C., Humbert, N. & Ward, T. R. (2005) Artificial metalloenzymes based on the biotin-avidin technology for the enantioselective reduction of ketones via transfer-hydrogenation, *Proc Natl Acad Sci USA* **102**, 4683–7.

12. Zocchi, A., Humbert, N., Berta, T. & Ward, T. R. (2003) Efficient expression and mutation of avidin and streptavidin as host proteins for enantioselective catalysis, *Chimia* **57**, 589–92.

13. Reetz, M. T. (2004) Controlling the enantioselectivity of enzymes by directed evolution: practical and theoretical ramifications, *Proc Natl Acad Sci USA* **101**, 5716–22.

14. Reetz, M. T. (2001) Combinatorial and evolution-based methods in the creation of enantioselective catalysts, *Angew. Chem. Int. Ed. Engl.* **40**, 284–310.

15. Kada, G., Falk, H. & Gruber, H. J. (1999) Accurate measurement of avidin and streptavidin in crude biofluids with a new, optimized biotin-fluorescein conjugate, *Biochim. Biophys. Acta.* **1427**, 33–43.

16. Humbert, N., Zocchi, A. & Ward, T. R. (2005) Electrophoretic behavior of streptavidin complexed to a biotinylated probe: a functional screening assay for biotin-binding proteins, *Electrophoresis* **26**, 47–52.

17. Sambrock, J. & Russel, D. W. *Molecular Cloning. A Laboratory Manual*, Cold Spring Harbor Laboratory Press, Plainview, NY, 2001.

18. Laemmli, U. K. (1970) Cleavage of structural proteins during the assembly of the head of bacteriophage T4, *Nature* **227**, 680–5.

19. Green, N. M. (1975) Avidin, *Adv. Protein Chem.* **29**, 85–133.

20. Freitag, S., Le Trong, I., Chilkoti, A., Klumb, L. A., Stayton, P. S. & Stenkamp, R. E. (1998) Structural studies of binding site tryptophan mutants in the high-affinity streptavidin-biotin complex, *J. Mol. Biol.* **279**, 211–21.

7

Application of Biotin-4-Fluorescein in Homogeneous Fluorescence Assays for Avidin, Streptavidin, and Biotin or Biotin Derivatives

Andreas Ebner, Markus Marek, Karl Kaiser, Gerald Kada, Christoph D. Hahn, Bernd Lackner, and Hermann J. Gruber

Summary

Biotin-4-fluorescein (B4F) is a convenient molecular probe for (strept)avidin and for unlabeled biotin in homogeneous fluorescence assays. The primary standard is a 16 µM working solution of d-biotin which is used to titrate an aliquot of a (strept)avidin stock solution while monitoring the tryptophane fluorescence of (strept)avidin. This serves to standardize the (strept)avidin stock solution, an aliquot of which is then titrated with a roughly 16 µM working solution of B4F while monitoring the fluorescence of B4F. Specific binding is accompanied by quenching, but after saturation of all binding sites, the appearance of free ligand causes a sharp rise of intense fluorescence, the beginning of which allows to calculate the effective concentration of B4F in the working solution. Measurement of avidin in a crude sample is exemplified by mixing 8 pmol of B4F with various amounts of diluted egg white in a volume of 1 mL. Hereby, the extent of fluorescence quenching linearly correlates with the concentration of functional avidin. Moreover, a sharp minimum of fluorescence is observed when exactly 2 pmol of avidin is present in the sample. The latter assay has been adapted to measure between 0.5 and 5 pmol of d-biotin in 1 mL of sample by adding 1.9 pmol of avidin and 8 pmol of B4F. This competitive assay correctly measures the small dose of d-biotin in multivitamin tablets (e.g., 150 µg in 5 g of solid) after subtracting the background fluorescence of the colored aqueous solution.

Key Words: Biotin; avidin; streptavidin; fluorescence; fluorescein; titration.

From: *Methods in Molecular Biology, vol. 418: Avidin-Biotin Interactions, Methods and Applications*
Edited by: R. J. McMahon © Humana Press, Totowa, NJ

1. Introduction

"Fluorescent biotins" are biotin-spacer-fluorophore conjugates which fall into three categories. (i) Fluorescent biotins with poly(ethylene glycol) spacers retain high fluorescence even after binding to avidin or streptavidin *(1,2)*. (ii) Fluorescent biotins with shorter spacers usually suffer from quenching after binding to (strept)avidin, and the quenching effect can be used for biospecific detection of (strept)avidin in the fluorimeter or fluorescence reader *(1–5)*. Nevertheless, the residual fluorescence in the bound state can be high enough for detection in heterogeneous fluorescence assays, as exemplified by prelabeling of new streptavidin mutants with biotin-4-fluorescein (B4F), followed by gel electrophoresis in 0.1% sodium dodecyl sulfate (SDS) and subsequent visualization of the streptavidin band on a transilluminator *(6)*. (iii) Some short biotin–fluorophore conjugates, such as biotin-4-Cy3 and biotin-4-Cy5 *(7)*, as well as lucifer yellow biocytin and cascade blue biotin (G. Kada, unpublished results) fluoresce well, even in the bound state, and are therefore good markers for (strept)avidin in heterogeneous assays.

The present study exemplifies the application of B4F in homogeneous bioanalytical assays for (strept)avidin or for unlabeled biotin in complex, colored samples such as egg white or commercial multivitamin preparations, respectively. The primary standard is a 16 µM working solution of d-biotin. The first task is to titrate a defined aliquot of an approximately 2 µM avidin or streptavidin working solution with exactly 16 µM d-biotin which yields the effective concentration of (strept)avidin. Immediately thereafter, an equal aliquot of the same (strept)avidin working solution is titrated with a 10–16 µM working solution of B4F, yielding the accurate functional concentration of B4F. All aliquots of the same batch of B4F are labeled with this effective concentration and stored for up to 1 month at –20°C or for years at –70°C. Aliquots from the standardized B4F are used to measure avidin or streptavidin in samples which may contain a large excess of other proteins (e.g., 10^3-fold to 10^4-fold as in egg white). Typically, 8 pmol of biotin is added to a sample containing <2 pmol of (strept)avidin, and the concentration of the latter is calculated from fluorescence quenching, as compared to absence of avidin. Unlabeled biotin or biotin derivatives (0.5–5 pmol) can also be measured in 1 mL of an unknown sample by adding 1.9 pmol of avidin and 8 pmol of B4F and comparing the fluorescence to a calibration curve with known biotin.

2. Materials

2.1. Stock Solutions and Working Solutions of d-Biotin, Avidin, and Streptavidin

1. Buffer A: 100 mM NaCl, 50 mM NaH_2PO_4, and 1 mM ethylenediaminetetraacetic acid (EDTA), pH adjusted to 7.5 with NaOH.

2. Buffer B: Fatty acid-free bovine serum albumin (BSA) (e.g., Roche, 775827) is dissolved at 10 mg/mL in buffer A (e.g., 100 mg in 10 mL) and 1 mL aliquots are stored at –20°C. Before use, a 1 mL aliquot is diluted with 99 mL buffer A to obtain buffer B.

3. D-biotin stock solution (4 mM = 0.997 mg/mL, 244.3 g/mol) in buffer A: D-biotin (somewhat less than 99.7 mg) is weighed into a 100-mL graded cylinder. The final volume V of the 4 mM d-biotin stock solution is calculated from the actual mass of d-biotin (M, in mg): $V = M/0.997$. The dry d-biotin is dissolved in buffer A at the final volume of V (in mL). Aliquots (e.g., 1 mL) are stored at –20°C for up to 1 month or at –70°C for years.

4. D-biotin standard solution (16 µM): One aliquot of the frozen 4 mM d-biotin stock solution is quickly warmed in approximately 25°C water and mixed before opening. Hundred microliters of 4 mM d-biotin is carefully diluted to 25 mL with buffer A, and 500 µL aliquots are frozen in 500-µL Eppendorf tubes for up to 1 month at –20°C. Immediately before use, an aliquot is thawed in approximately 25°C water, carefully mixed before opening, and immediately consumed or discarded.

5. Avidin stock solution (~5 µM, 66,000 g/mol) and working solution (2 µM): Affinity-purified avidin from Sigma (A-9275) was used here, but pure avidin from other sources is also suitable. Between 3 and 5 mg of solid is weighed into a glass vial and dissolved in 10 mL of buffer A. The UV-Vis spectrum is measured and A_{282} (per cm) is divided by the molar absorptivity of avidin ($\varepsilon_{282} = 96,000/M/cm$) to calculate the molar concentration. In a typical example, 4.2 mg of dry solid (batch specification 10.8 units/mg of solid and 11.9 units/mg protein, i.e., 83% of the theoretical activity of 14.4 units/mg avidin) dissolved in 10 mL buffer A showed $A_{282} = 0.567$ per cm, indicating a protein content of 5.8 µM. Taking into account the 83% functional purity (see above), this corresponds to a functional avidin concentration of 4.8 µM. This solution is diluted with buffer A (e.g., by a factor of 4.8 µM/2 µM = 2.4 in our example) in order to obtain a working solution with 2 µM functional avidin according to batch specifications.

6. Streptavidin stock solution (~20 µM, 60,000 g/mol) and working solution (2 µM): The purchased quantity of 1 mg (16.7 nmol, Sigma, 85878) is dissolved in 833 µL of buffer A. Aliquots (exactly 200 µL each, ~4 nmol) are distributed in 2-mL Eppendorf vials, rapidly frozen in liquid nitrogen, and stored at –20°C or –70°C. Before use, one aliquot is thawed, shortly centrifuged to collect all liquid at the bottom, and mixed with 1.8 mL of buffer A. If desired, the molar concentration can be checked by measuring the UV absorbance A_{282} against buffer A and dividing by the molar absorptivity (224,000/M/cm).

2.2. Stock Solution and Working Solution of B4F

1. Dry aliquots of B4F (~8 µmol each): 5 mg B4F (78 µmol, MW = 344.7, Molecular Probes, B10570) is dissolved in 2 mL ethanol, and 0.2 mL aliquots are dispensed in small threaded glass vials with a Hamilton syringe. The solvent is blown off with argon or nitrogen gas. The vials are evacuated in a round-bottomed flask

for 1–2 h, filled with argon or nitrogen gas, screw-capped, sealed with Parafilm, and stored at –20°C or –70°C over desiccant (e.g., blue gel).

2. Dimethyl sulfoxide (DMSO) stock solution of B4F ($\sim$500 μM): One dry portion of B4F ($\sim$0.5 mg, $\sim$8 μmol) is dissolved in 1.5 mL of DMSO (analytical grade or better) and stored at –20°C for <1 month or at –70°C for years.

3. Buffer C: 100 mM NaCl, and 50 mM H_3BO_3, pH adjusted to 9.0 with NaOH. Buffer C (pH 9) is solely used for estimating the molar concentration of a B4F stock solution from the UV-Vis absorption spectrum because at pH 9 the molar absorptivity ε_{494} is known to be 68,000/M/cm. Please note that buffer C is inadequate for any experiments with avidin or streptavidin.

4. Estimation of B4F concentration in the DMSO stock solution: 40 μL of the DMSO stock solution is mixed into 1960 μL buffer C (not buffer A), and the absorption spectrum is measured from 400 to 600 nm in a cuvette with 1 cm light path. A_{494} (per cm) is divided by the molar absorptivity (ε_{494} = 68,000/M/cm) to give the nominal concentration of B4Flu. In a typical experiment, A_{494} = 0.859 per cm was measured, division by 68,000/M/cm gave a nominal concentration of 12.63 μM in the cuvette and of 12.63 μM × (2000 μL/40 μL) = 632 μM in the DMSO stock solution.

5. Working solution of B4F ($\sim$16 μM in buffer A): Above, an aliquot of the DMSO stock solution has been diluted with buffer C (not buffer A) to estimate the approximate concentration of fluorescent biotin by UV-Vis absorption. Now, a proper volume of the same DMSO stock solution is made up to 5 mL with buffer A (not buffer C) to obtain a $\sim$16 μM working solution. In our case (*see* **step 4**), 5 mL × (16 μM/632 μM) = 0.127 mL of DMSO stock solution was carefully pipetted and rinsed into <4.8 mL buffer A, made up to 5 mL with buffer A and carefully mixed. Five hundred microliter aliquots are quickly dispensed into 500-μL Eppendorf tubes and stored at –20°C for up to 1 month or at –70°C for years. After thawing in 25°C water, the solution is well mixed before opening and consumed within one workday.

3. Methods

3.1. Standardization of Avidin or Streptavidin Working Solutions by Titration with d-Biotin and Monitoring of Tryptophane Quenching

1. The biotin-binding pockets of avidin and streptavidin contain a tryptophan residue, the fluorescence of which is quenched upon binding of biotin. Although avidin and streptavidin contain additional tryptophan residues in each subunit, the total fluorescence is quenched by 40% *(8)* or 30% *(3)*, respectively, when four biotins are bound per tetramer. Titration of (strept)avidin with a standardized solution of d-biotin leads to a linear reduction of fluorescence up to the point at which all biotin-binding sites have been filled with biotin (*see* **Fig. 1**). Further addition of d-biotin does not alter fluorescence, except for the very minor dilution effect and slow bleaching of tryptophan by the 290-nm irradiation. The consumption

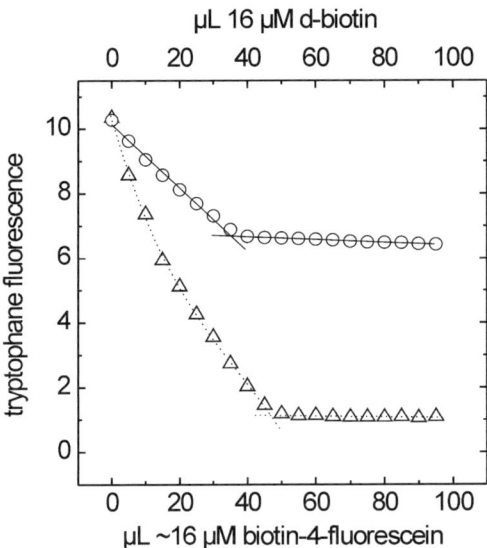

Fig. 1. Titration of avidin with d-biotin or biotin-4-fluorescein (B4F) in a stirred cuvette while monitoring tryptophane fluorescence. Buffer A, 1920 µL, was mixed with 80 µL of approximately 2 µM avidin and stirred while adding 5 µL portions of exactly 16 µM d-biotin (circles, top axis) or approximately 16 µM B4F (triangles) from a Hamilton syringe at 1-min time intervals. Consumption of 35.2 µL 16 µM d-biotin (circles) indicated the presence of 563 pmol binding sites or 141 pmol avidin tetramer in the cuvette. The same amount of avidin consumed 52.1 µL of B4F working solution (triangles, bottom axis), indicating an effective concentration of 10.8 µM. The intercept between the initial nonlinear decrease and the constant fluorescence at high amounts of B4F is poorly defined which explains the deviation from the 12.0 µM concentration value obtained when monitoring fluorescence at 525 nm (*see* circles in **Fig. 2**).

of biotin up to the breakpoint between the two phases of the titration, therefore, allows to calculate the number of biotin-binding sites in the cuvette contents.

2. One 500 µL aliquot of the 16 µM d-biotin standard solution is quickly warmed to room temperature in 25°C water and thoroughly mixed before opening.

3. Fluorimeter settings: 290-nm excitation (slit 5 nm), 340-nm emission (slit 10 nm). The vertical scaling factor (or photomultiplier voltage) must be adjusted so as to obtain a noise-free fluorescence signal from avidin or streptavidin before biotin is being added (see below).

4. The cuvette is filled with 1920 µL buffer A, and exactly 80 µL of the approximately 2 µM avidin (or streptavidin) working solution is pipetted into the stirred buffer, taking care to rinse all of the protein from the pipette tip into the buffer.

5. A gas-tight Hamilton syringe (100 µL) is filled with 100 µL of 16 µM d-biotin standard solution. Air bubbles are removed by repeated ejection into the biotin solution.

6. Fluorescence is read ($t = 0$ min) and 5 µL of 16 µM d-biotin is added with stirring. At $t = 1$ min, fluorescence is read again and the next 5 µL aliquot is added with stirring. This step is repeated up to a total added volume of 95 µL biotin addition.

7. The fluorescence data are plotted versus d-biotin consumption, and the two linear periods are fitted by straight lines, as shown in **Fig. 1**. The number of biotin-binding sites in the cuvette contents is calculated from the biotin consumption at the intersection of the two lines. In **Fig. 1**, the breakpoint indicated a consumption of 35.2 µL × 16 µM = 563 pmol of d-biotin for saturation of all specific binding sites in avidin. This corresponds to 563/4 = 141 pmol of functional avidin tetramers in the cuvette, as well as in the 80 µL of avidin stock solution used to prepare the cuvette contents. Thus, the "functional" concentration of avidin tetramers in approximately 2 µM stock solution is 141 pmol/80 µL = 1.76 µM.

3.2. Titration of a Known Avidin Standard with an Unknown Working Solution of B4F While Monitoring Fluorescence at 525 nm

1. Above, a defined aliquot of approximately 2 µM avidin has been titrated with an exactly 16 µM d-biotin standard to measure the functional concentration of avidin. Now, an equal aliquot of the same approximately 2 µM avidin solution is titrated with approximately 16 µM B4F while monitoring the fluorescence of the ligand at 525 nm. As can be seen from **Fig. 2** (circles), specific binding of B4F to avidin is accompanied by strong fluorescence quenching. Quenching is most pronounced when all four binding sites in the avidin tetramer are occupied by B4F (e.g., at $x = 47.1$ µL, *see* **Fig. 2**), followed by a sharp linear rise in fluorescence when excess of B4F is added. The consumption of B4F at the breakpoint reflects stoichiometric binding of B4F to the known amount of avidin, thus the effective concentration of B4F in the nominally approximately 16 µM working solution is obtained from this type of titration.

2. Fluorimeter settings: 490-nm excitation (5-nm slit), 525-nm emission (5-nm slit).

3. The cuvette is filled with 1920 µL buffer A, and exactly 80 µL of the approximately 2 µM avidin working solution is pipetted into the stirred buffer, taking care to rinse all of the protein from the pipette tip into the buffer solution. In **Fig. 2** (circles), the accurate amount of avidin is known from **Fig. 1** (circles) to be 80 µL × 1.76 µM = 141 pmol, corresponding to 563 pmol binding sites. Fluorescence is read ($t = 0$ min), and 5 µL aliquots of approximately 16 µM B4F are added from a gas-tight Hamilton syringe with stirring at 1-min intervals and reading the fluorescence "before" each successive addition, in close analogy to **Subheading 3.1.**

4. The fluorescence data are plotted versus biotin consumption (*see* circles in **Fig. 2**). The curved segment between 0 and 40 µL is fitted to $y = a + bx + cx^2$, while the data points with $x > 50$ µL are fitted to a straight line. The breakpoint at 47.1 µL means that this volume of the B4F working solution contained the 563 pmol of ligand necessary to saturate the 141 pmol of avidin in the cuvette (*see* **step 3**). Thus, the concentration of B4F in the working solution is 563 pmol/47.1 µL = 12.0 µM. The deviation from approximately 16 µM concentration estimated

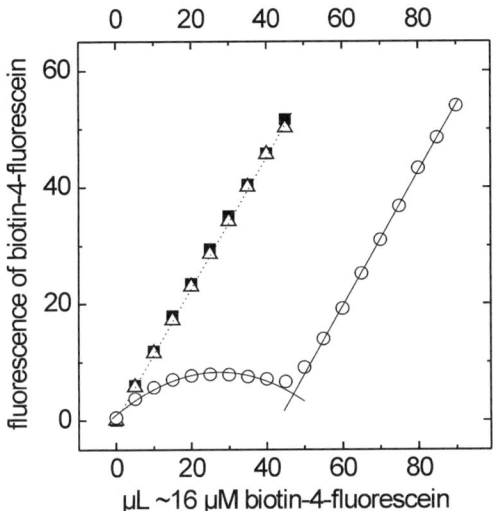

Fig. 2. Titration of avidin with biotin-4-fluorescein (B4F) in a stirred cuvette while monitoring emission of B4F. Buffer A, 1920 µL, was mixed with 80 µL of 1.76 µM avidin (563 pmol binding sites, *see* **Fig. 1**) and stirred while adding 5 µL portions of approximately 16 µM B4F (circles) from a Hamilton syringe at 1-min time intervals. Consumption of 47.1 µL B4F at the breakpoint (circles) indicates a concentration of 563 pmol/47.1 µL = 12.0 µM in the working solution of B4F. Parallel control experiments with the same amount of avidin but pre-blocked with 10 µL 4 mM biotin (open triangles) or without any avidin (filled squares) indicated that binding of B4F is fully specific.

from UV-Vis absorbance (*see* **Subheading 2.2.**) must be attributed to the fact that B4F has an actual molar absorptivity which significantly differs from the nominal value 68,000/M/cm at pH 9. The alternate explanation, that is, presence of free fluorescein without biotin residues, is less likely because parallel thin layer chromatography of B4F indicated high purity (not shown). Moreover, the same discrepancy has been observed before *(4)*.

3.3. Estimation of Avidin in Complex Biofluids by Titration with B4F ("Forward Titration")

1. **Figure 2** shows that the fluorescence of B4F reports on its binding to specific binding sites in avidin. In **Fig. 3**, the same titration protocol is used to specifically detect avidin in a 10-fold dilution of egg white although ≪0.01% of total protein consists of avidin.
2. The egg is warmed to room temperature, the egg white is carefully separated from the yolk and put into a graded glass beaker (100 mL nominal volume). Egg white

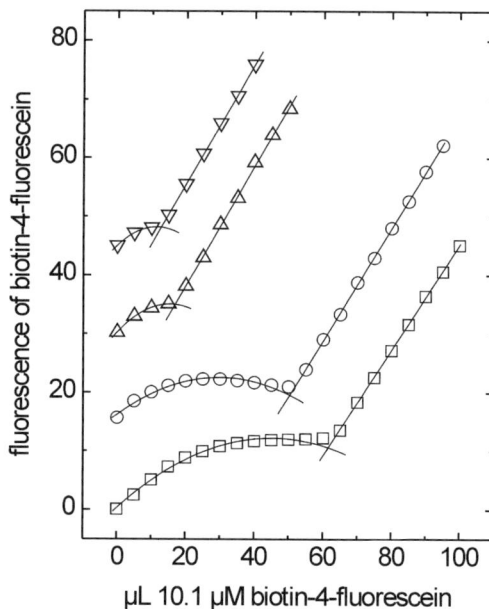

Fig. 3. Measurement of avidin in egg white with biotin-4-fluorescein (B4F). In each experiment, a 2 mL aliquot of 10-fold diluted egg white was titrated with 5 μL increments of 10.1 μM B4F, in close analogy to **Fig. 2**, except that another batch of B4F working reagent was used. The curves for egg no. 1, 2, and 3 (numbered from top to bottom) were vertically displaced by multiples of 15 fluorescence units to avoid data overlap, only the curve for egg no. 4 is shown as measured. The breakpoints between the curved segments and the straight lines were determined as 12.3, 16.9, 29.5, and 61.5 μL, corresponding to functional avidin contents of 31, 43, 125, and 155 pmol in 2 mL of 10-fold diluted egg white, respectively. The total avidin and protein contents of each egg is shown in **Table 1**.

contaminated with yolk must be discarded because the latter contains much d-biotin. The volume is recorded and the egg white is stirred with a magnetic spin bar without causing bubble formation. After 2 min, an equal volume of buffer is slowly added within 3 min. Ten milliliters of the mixture is transferred into a graded cylinder and diluted to 50 mL with buffer A.

3. In the fluorimeter cuvette, 2 mL of the 10-fold diluted egg white is stirred and titrated with successive 5 μL aliquots from a approximately 16 μM B4F working solution at 1-min intervals, as described in **Subheading 3.2.** for a sample of pure avidin.

4. **Figure 3** shows the results obtained with four different eggs. The bottom curve is shown as measured while the other curves have been vertically displaced by 15, 30, and 45 fluorescence units to avoid overlap. The bottom curve shows the highest avidin contents. The next curve indicates less avidin but it shows more

quenching in the breakpoint region than any other curve. This means that this egg white was least contaminated with biotin while in the other egg whites avidin had been partially occupied with endogenous biotin.

3.4. Titration of a Known Amount of B4F with an Unknown Sample of Avidin ("Reverse Titration")

1. Gradual titration of avidin with B4F ("forward titration," **Subheading 3.3.**) is accompanied by a *nonlinear* fluorescence response (*see* **Figs 2 and 3**) because fluorescence quenching is much stronger when 3–4 rather than 1–2 ligands are bound per avidin tetramer *(4)*. In contrast, a *linear* dose response is observed in "reverse titration," that is, when, for instance, ≤ 2 pmol of avidin tetramer is added to a fixed amount of 8 pmol B4F. Under such conditions, B4F is either unbound or it takes part in a 4:1 complex between B4F and (streptavidin, thus the average fluorescence of the fluorescent biotins linearly correlates with the amount of avidin added (at $0 < x < 50$ µL in **Fig. 4A**). Fluorescence quenching is maximal when 2 pmol of avidin are mixed with 8 pmol of B4F (at $x = 40$ µL in **Fig. 4A**). Higher amounts of avidin, however, result in partial fluorescence recovery because now less than 4 ligands are bound per avidin tetramer in which case quenching is weaker (at $x > 40$ µL in **Fig. 4A**). The same dose responses have been reported for streptavidin *(5)* and for various derivatized forms of avidin or streptavidin *(6)*.
2. Fortunately, a linear dose response of 8 pmol B4F is also observed when crude egg white with <2 pmol of avidin is added to 8 pmol of B4F (*see* **Fig. 4B**), thus "reverse titration" is applicable to crude biological samples also.
3. Fluorimeter settings: 490-nm excitation wavelength (10-nm slit), 525-nm emission wavelength (10-nm slit).
4. Twenty-five milliliters of 160 nM B4F reagent is prepared by dilution of an aliquot from the approximately 16 µM working solution that has been standardized as shown in **Fig. 2** (circles). In our test series, 333 µL of 12.0 µM B4F was made up to 25 mL with buffer A.
5. Twenty-five milliliters of 50 nM avidin reagent is prepared by mixing buffer B (not buffer A) with an aliquot of the approximately 2 µM avidin working solution that has been standardized as shown in **Fig. 1** (circles). In our example, 710 µL of 1.76 µM avidin was made up to a volume of 25 mL with buffer B.
6. For a typical calibration curve, $n \times 17$ polystyrene tubes (for n replicates, 4 mL tube volume) are filled with $1000 - x$ µL of buffer B (not buffer A) plus x µL of 50 nM avidin reagent ($x = 0, 5, 10, \ldots 80$). To each tube, 50 µL of 160 nM B4F is added from an Eppendorf Multipette while vortexing, followed by 3 short pulses of vortexing, 10 min of incubation, and reading of fluorescence in a small cuvette, taking care to let the photomultiplier recover from shut-down each time after lid closure.
7. The calibration curve for known avidin samples is shown in **Fig. 4A**. The intercept between the two linear branches is located at $x = 39.0$ µL of 50 nM avidin,

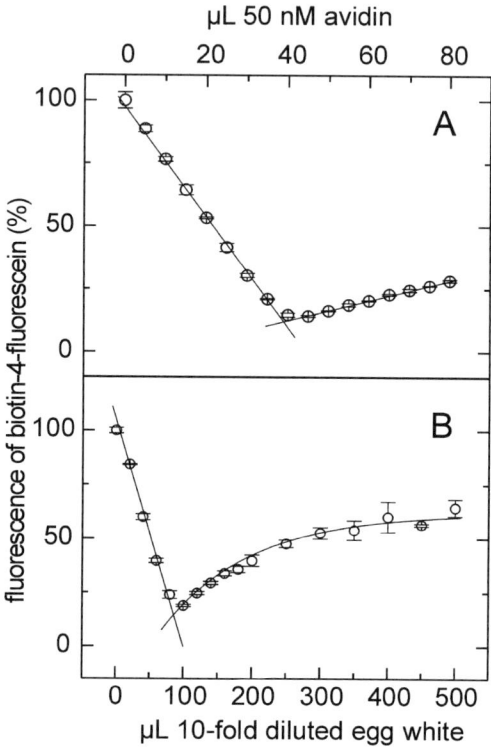

Fig. 4. Quenching of biotin-4-fluorescein (B4F) fluorescence by addition of various amounts of pure avidin (**A**) or of diluted egg white (**B**). For each data point, a 1 mL sample was prepared by mixing the indicated volume of 50 nM avidin (**A**) or of 10-fold diluted egg white (**B**)with buffer B, followed by addition of 8 pmol of B4F (50 µL of a 160 nM solution). Fluorescence was read after 10-min incubation time and renormalized to 100% fluorescence in the absence of avidin or egg white.

corresponding to an apparent avidin content of 1.95 pmol at the point when all available binding sites were filled with B4F ligands. For practical purposes, and because 1.95 pmol of avidin calculated at the point of binding site saturation closely agrees with the expected value of 2.00 pmol, the concentration of the 50 nM avidin reagent is corrected to 50 nM $\times$ (2.00/1.95) = 51.3 nM. In a more rigorous approach, the 50 nM avidin reagent concentration would remain unchanged while the 160 nM B4F reagent would be redefined as 156 nM, as well as the 12.0 µM B4F working solution as 11.7 µM. The reason is that the 50 nM avidin reagent had been prepared from the 1.76 µM avidin working solution that had been standardized by unequivocal titration with d-biotin (*see* **Fig. 1**, circles), while the standardization of B4F against avidin suffers from slight ambiguities in determining the point of binding site saturation (*see* circles in **Fig. 2**).

8. The egg white series in **Fig. 4B** was prepared in the same way as the calibration series in **Fig. 4A**, except that the indicated volumes (see *x* axis) of 10-fold diluted egg white (*see* **Subheading 3.3.**) were made up to 1 mL with buffer B before the 50 μL portions of 160 nM B4F were added. Interestingly, the clear-cut titration curve in **Fig. 4B** was obtained from the egg white having the lowest avidin content (*see* top curve in **Fig. 4A**). The breakpoint between the linear fall and the nonlinear rise in **Fig. 4B** was determined at *x* = 86.1 μL of 10-fold diluted egg white. At this point, 8 pmol of B4F was bound by exactly 2 pmol of avidin, thus the avidin concentration in the 10-fold diluted egg white was 2 pmol/86.1 μL = 23.2 μM, corresponding to 10.2 nmol (0.68 mg) of avidin in 44 mL of undiluted egg white for egg no. 1 (*see* **Table 1**). The discrepancy is explained by partial denaturation (*see* **Notes 1–9**) at low avidin concentration in a stirred cuvette (*see* **Fig. 3**, **Table** 1) which is absent in unstirred plastic tubes (*see* **Fig. 4B**).

3.5. Measurement of Unlabeled Biotin in Complex Samples by Competition with B4F for the Binding Sites in Avidin

1. **Figure 4A** shows that the fluorescence of 8 pmol B4F is reduced by 88% when mixed with 2 pmol of avidin. Less avidin causes less quenching, and the dose response is linear. This implies that 2 pmol of avidin partially pre-blocked with unlabeled biotin should also cause less quenching than 2 pmol of unoccupied avidin. Working with 8 pmol of B4F and 2 pmol of avidin, however, is too critical because pipetting errors would lead to samples with more binding sites than necessary to bind the 8 pmol of B4F, thus a small quantity of unlabeled biotin could be bound without affecting B4F fluorescence. In practice, therefore, 1 mL samples with known (or unknown) biotin contents are equilibrated with 1.9 pmol (not 2 pmol) avidin tetramer, followed by addition of 8 pmol B4F. The result is a largely linear dose response of B4F fluorescence, both for known biotin standards (*see* solid circles in **Fig. 5**) and for dilute samples from a commercial multivitamin preparation according to the stated content of d-biotin (*see* open

Table 1
Avidin and Protein Content of the Egg White Samples Shown in Fig. 3 (From Top to Bottom)

Egg no.	Egg white volume (ml)	Avidin content	Total protein (g)
1	44	6.8 nmol (0.43 mg)	4.3
2	34	7.2 nmol (0.45 mg)	6.1
3	36	23 nmol (1.5 mg)	4.2
4	32	25 nmol (1.7 mg)	4.3

The protein content was estimated from the UV spectrum (A_{280}) of a 200-fold dilution of egg white with buffer A, using ovalbumin as a relative standard (A_{280} = 7.35 at 1% w/v) (**10**).

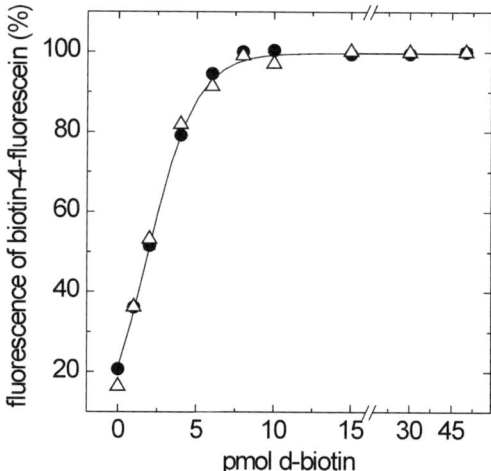

Fig. 5. Homogeneous fluorescence assay for unlabeled biotin. One milliliter samples of buffer B with the indicated content of d-biotin (solid circles) were mixed with 1.9 pmol of avidin (in 50 μL), incubated for 10 min, mixed with 8 pmol of biotin-4-fluorescein (in 50 μL), and incubated for another 10 min before reading fluorescence. The fluorescence at the highest biotin concentration was renormalized to 100%. The open triangles were measured from a commercial multivitamin tablet, and they are identical to the open triangles in **Fig. 6**, except that the maximal fluorescence at high biotin concentration was renormalized to 100% in this **Fig. 5**. All data points represent the results of triplicates, but the standard deviations were smaller than the symbol sizes and were omitted from the graph for clarity.

triangles in **Fig. 5**), taking into account endogenous fluorescence of the multivitamin solution (*see* **Fig. 6** and below).

2. d-Biotin, 100 nm, is freshly prepared from 156 μL of a 16 μM d-biotin standard solution (freshly thawed in ~25°C water) plus buffer B (not buffer A) in a final volume of 25 mL.

3. Avidin, 38 nM, is prepared by proper dilution of the 50 nM avidin stock solution (*see* **Subheading 3.4.**) with buffer B (not buffer A). Based on the breakpoint in **Fig. 4A**, the 50 nM avidin concentration had been corrected to 51.3 nM (*see* **Subheading 3.4.**). Therefore, the 38 nM avidin solution was prepared by diluting 18.5 mL 51.3 nM avidin reagent to 25 mL with buffer B.

4. The 160 nM B4F reagent is identical to that described in **Subheading 3.4.** (*see* **Fig. 4A and B**).

5. For the calibration curve (*see* closed circles in **Fig. 5**), 1000 − x μL of buffer B (not buffer A) and *x* μL of 100 nM d-biotin (*x* = 0, 10, 20, 40, 60, 80, 100, 150, 300, and 500, optionally in triplicates) are mixed in 4-mL polystyrene tubes, and a 50 μL portion each of 38 nM avidin is added from an Eppendorf Multipette while vortexing, followed by 3 short pulses of vortexing. After 10 min of equilibration, a

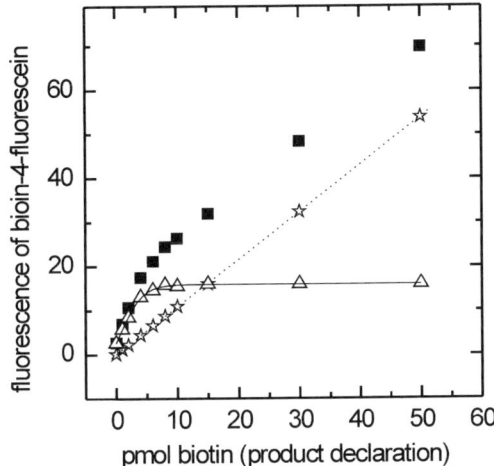

Fig. 6. Measurement of biotin in a commercial multivitamin tablet. One effervescent tablet (5 g with a declared biotin content of 150 µg) was dissolved in water (pH adjusted to ~8) and diluted to 100 nM nominal biotin concentration with buffer B. Aliquots were further mixed with buffer B to prepare the 1 mL test samples with the indicated nominal biotin content. After addition of 1.9 pmol avidin (50 µL, 10-min incubation) and 8 pmol biotin-4-fluorescein (B4F) (50 µL, 10-min incubation), fluorescence was read and plotted versus nominal biotin content (solid squares). In a parallel control series (open stars), buffer A (100 µL) was added instead of avidin and B4F. Subtraction of endogenous sample fluorescence (open stars) from sample fluorescence (solid squares) gave the fluorescence contribution of B4F (open triangles). The data points represent the results of triplicates, but the standard deviations were smaller than the symbol sizes and were omitted in the graph for clarity.

50 µL portion of 160 nM B4F is added to each tube from an Eppendorf Multipette while vortexing, followed by 3 short pulses of vortexing and 10 min of incubation. Fluorescence is read with the same settings as in **Subheading 3.4.** and plotted versus free biotin concentration (*see* closed circles in **Fig. 5**.

6. For measurement of d-biotin in a multivitamin effervescent tablet, one 5-g tablet (Berocca) with a stated biotin content of 150 µg (613 nmol) was allowed to dissolve in 60 mL of water, followed by pH adjustment to approximately 8 with NaOH in order to ensure full solubility of biotin. The solution was transferred to a graded cylinder and diluted to 76.8 mL with water, resulting in a nominal d-biotin concentration of 613 nmol/76.8 mL = 8 µM. Six hundred and twenty-five microliters of this solution was made up to 50 mL with buffer B, resulting in a vitamin solution with a nominal biotin concentration of 100 nM. All further steps were the same as described in step 5 for the known 100 nM d-biotin standard. The fluorescence data were plotted versus biotin content according to the product declaration (*see* closed squares in **Fig. 6**). The linear slope at

high biotin concentrations was obviously due to endogenous fluorescence of the vitamin solution in absence of B4F. This was confirmed in a repetition of the same tube series in which 100 μL buffer A was added instead of 50 μL avidin and 50 μL of B4F (*see* open stars in **Fig. 6**). The difference between the two curves (squares and stars) gave the fluorescence contribution of B4F in each sample. It was renormalized to 100% fluorescence at the maximal biotin content and plotted in **Fig. 5** (open triangles). The perfect agreement with the calibration curve (*see* solid circles in **Fig. 5**) clearly shows that the tablet contained exactly the stated amount of d-biotin and that the correction method for endogenous sample fluorescence shown in **Fig. 6** is valid.

4. Notes

1. EDTA in buffer A is optional and can be omitted, if desired. The high buffering capacity of 50 mM phosphate is necessary to quickly dissolve d-biotin when preparing the 4 mM stock solution. For all other purposes, buffer A can be replaced by phosphate-buffered saline or similar.
2. Sodium azide (5 mM) may be used only to preserve the stock solutions of avidin and streptavidin. Azide should be avoided in the measuring buffer and is strictly prohibited in solutions of d-biotin and B4F as it greatly accelerates biotin degradation by an unknown mechanism.
3. Fluorescein is a particularly light-sensitive fluorophore, therefore, bright light is avoided and aluminum foil is used to protect samples of B4F from light.
4. For minimization of bleaching (of tryptophane and of B4F) in the fluorimeter, the shutter is kept closed until the start of the fluorescence titration. Bleaching can further be lowered when reducing the excitation slit which results in a noisier signal and can partially be compensated for by a longer integration time.
5. When titrating (strept)avidin with d-biotin, cumulative titration of one avidin sample in a stirred cuvette (*see* circles in **Fig. 1**) is much preferred because noncumulative titration (in a series of tubes with different biotin/avidin ratios) is tedious and the data scatter more than explained by imprecisions of pipetting. The obvious reason is that tryptophane fluorescence is relatively weak, thus the measurement is sensitive to environmental dust (even with filtered buffers). The amount of dust is constant in cumulative titration but variable in noncumulative titration. As expected, no such discrepancy in the data quality of cumulative versus noncumulative titration is seen when titrating (strept)avidin with B4F and measuring fluorescence at long wavelength (*see* circles in **Fig. 2**, "forward titration"). Nevertheless, cumulative titration in the stirred cuvette is preferred because of much shorter assay time and because the gas-tight Hamilton syringe serves as a burette which ensures high volumetric accuracy even with untrained personnel.
6. The procedures for titration of streptavidin with d-biotin of B4F are the same as for avidin, and the data are very similar *(3,4)*.

7. The results from the cumulative titration of avidin with d-biotin (*see* **Fig. 1**) were confirmed by noncumulative titration in two series with 17 polystyrene tubes each: A variable volume of buffer A plus 25 µL of the same approximately 2 µM stock solution and a variable volume (0, 25, 50, ... 400 µL) of exactly 1 µM d-biotin (prepared by diluting 16 µM d-biotin with buffer A) were mixed at a final volume of 1 mL. After 10 min, each sample was poured in a small cuvette, and the fluorescence at 340 nm (10-nm slit) was read while exciting at 290 nm (10-nm slit). The plot of tryptophane fluorescence versus µL of 16 µM d-biotin (not shown) was similar to **Fig. 1** (circles). The breakpoint at 179 µL 1 µM d-biotin (179 pmol) indicated an effective avidin concentration of 44.8 pmol/25 µL = 1.79 µM in the approximately 2 µM avidin stock solution. The second series indicated a functional avidin concentration of 1.72 µM. Both values nicely confirm the 1.76 µM obtained in the cumulative titration shown in **Fig. 1**.

8. It should be mentioned that since the year 2003, no functional avidin has been detectable in chicken eggs purchased in local stores, the only exception being eggs from organic farms. The likely reason is the use of new animal feed with excessive contents of d-biotin.

9. The egg white in the top curve of **Fig. 3** was cross-checked with "reverse titration" (*see* **Fig. 4B**). The 1.5-fold avidin content found in the latter method (see **Subheading 3.3.** and **Fig. 4B**) indicates that cumulative titration in a stirred cuvette leads to significant denaturation at such low avidin contents, thus "reverse titration" in plastic tubes (*see* **Subheading 3.4.**) is the method of choice for measurement of avidin in dilute samples. In contrast, only 2.5% disagreement between cumulative "forward titration" (*see* **Fig. 2**) and noncumulative "reverse titration" (*see* **Fig. 4A**) was found because the avidin concentration in the stirred cuvette was high enough (>100 pmol/2 mL, as in **Fig. 2**). It can be summarized that stirring in the quartz cuvette causes denaturation at <100 pmol of avidin (in 2 mL of sample) which is not prevented by the high protein content of egg white, whereas long-term incubation and only short vortexing in plastic tubes does not cause any denaturation, even at very low (strept)avidin concentrations (as in **Figs 4–6**), provided that 0.1 mg/mL of BSA is included *(4,5)*.

Acknowledgment

This work was supported by the Austrian Science Foundation (FWF-P15295).

References

1. Gruber, H. J., Marek, M., Schindler, H., and Kaiser, K. (1997) Biotin-fluorophore conjugates with poly(ethylene glycol) spacers retain intense fluorescence after binding to avidin and streptavidin. *Bioconjug. Chem.* **8**, 552–559.

2. Marek, M., Kaiser, K., and Gruber, H. J. (1997) Biotin-pyrene conjugates with poly(ethylene glycol) spacers are convenient fluorescent probes for avidin and streptavidin. *Bioconjug. Chem.* **8**, 560–566.

3. Gruber, H. J., Kada, G., Marek, M., and Kaiser, K. (1998) Accurate titration of biotin-binding sites in avidin and streptavidin with biotin-fluorophore conjugates. *Biochim. Biophys. Acta* **1381**, 203–212.

4. Kada, G., Falk, H., and Gruber, H. J. (1999) Accurate measurement of avidin and streptavidin in crude biofluids with a new, optimized biotin-fluorescein conjugate. *Biochim. Biophys. Acta* **1427**, 33–43.

5. Kada, G., Kaiser, K., and Gruber, H. J. (1999) Rapid estimation of avidin and strep-tavidin by fluorescence quenching or fluorescence polarization. *Biochim. Biophys. Acta.* **1427**, 44–48.

6. Humbert, N., Zocchi, A., and Ward, T. R. (2005) Electrophoretic behaviour of streptavidin complexed to a biotinylated probe: a functional screening assay for biotin-binding proteins. *Electrophoresis* **26**, 47–52.

7. Gruber, H. J., Hahn, C. D., Kada, G., Riener, C, Harms, G. S., Ahrer, W., Dax, T. G., and Knaus, H.-G. (2000) Anomalous fluorescence enhancement of Cy3 and Cy3.5 versus anomalous fluorescence loss of Cy5 and Cy7 upon covalent linking to proteins and noncovalent binding to avidin. *Bioconjug. Chem.* **11**, 696–704.

8. Lin, H. J., and Kirsch, J. F. (1979) A rapid, sensitive fluorometric assay for avidin and biotin. *Methods Enzymol.* **62**, 287–289.

9. The Handbook – A Guide to Fluorescent Probes and Labeling Technologies, http://probes.invitrogen.com/handbook/sections/0400.html.

10. Kirschenbaum, D. M. (1975) Molar absorptivity and $A_{1\%}$ values for proteins at selected wavelengths of the ultraviolet and visible regions. *Anal. Biochem.* **64**, 186–213.

8

Quantitative Recovery of Biotinylated Proteins from Streptavidin-Based Affinity Chromatography Resins

Christoph Rösli, Jascha-N. Rybak, Dario Neri, and Giuliano Elia

Summary

The strong interaction between streptavidin and biotin is one of the most commonly exploited tools in chemistry and biology. Methods for the facile derivatization of a variety of molecules (in particular, proteins) with biotin have been introduced, in order to allow their efficient recovery, immobilization and detection with streptavidin-based reagents. However, when desired, the release of biotinylated proteins from the streptavidin-based reagents remains a major problem, due to the extraordinary stability of this complex. This chapter presents a protocol developed in our laboratory for the quantitative elution of biotinylated proteins from streptavidin sepharose, featuring harsh elution conditions and competition with free biotin. The usefulness of the method is shown by the recovery of biotinylated proteins from organ homogenates, obtained from mice perfused with a reactive ester derivative of biotin.

Key Words: Biotin; avidin; streptavidin; biotinylation; affinity chromatography.

1. Introduction

The high affinity and specificity of the interaction between avidin and vitamin H was discovered already in 1927, when it was observed that rats fed large quantities of egg white developed dermatitis, indicating malnutrition (*1*). Vitamin H, chemically identified as the biotin structure in 1940 (*2,3*), prevented this dermatitis. The malnutrition was eventually attributed to the depletion of biotin. Avidin, a protein present in the egg white, was complexing with the vitamin and interfering with the animal's nutrition. As early as 1941, the extraordinary affinity of avidin for biotin was recognized (*4*). Shortly thereafter

From: *Methods in Molecular Biology, vol. 418: Avidin-Biotin Interactions, Methods and Applications*
Edited by: R. J. McMahon © Humana Press, Totowa, NJ

in 1942, crystallization of avidin in a pure form was accomplished *(5)*. It was rapidly clear that the interaction of the relatively small vitamin H molecule with the egg white glycoprotein avidin and with the related bacterial protein streptavidin *(6,7)* could be easily converted in a affinity-based tool for several different purposes. Since then, chemical modification of a variety of molecules with biotin has been exploited as one of the most useful tools in biochemical and biomedical research. Biotinylated molecules (e.g., proteins, DNA and RNA) are easily detected with streptavidin derivatives [e.g., fluorophore, horseradish peroxidase (HRP) or alkaline phosphatase conjugates] or efficiently captured on streptavidin-coated solid supports (e.g., resins, magnetic beads, microtiter plates and chips). However, due to the very high stability of the biotin–streptavidin complex ($K_d = 10^{-15}$ M) *(8)*, the elution of biotinylated molecules from streptavidin-coated surfaces remains a major unsolved challenge.

Investigators have proposed many different approaches to overcome this problem and try to combine the ease of biotinylation reactions with suitable reversible capture/elution conditions, yet most of the methods available present drawbacks in certain experimental conditions. For example, biotinylation reagents have been introduced (e.g., sulfo-NHS-SS-biotin, Pierce, Rockford, MD, USA) that can be cleaved by reducing agents. Cleavable disulfide bridge-containing biotinylating reagents perform well, but may show insufficient stability in some biological fluids prior to purification. Proteins modified by means of photocleavable biotin derivatives (e.g., NHS-PC-LC-Biotin, Pierce) can be released using long-wave UV light *(9)*; yet, in some experimental conditions, it is not easy to achieve efficient illumination of the sample. Some resins on the market make use of modified (strept)avidin molecules with a lower affinity for biotin, therefore allowing elution under milder conditions [e.g., nitrated avidin derivatives *(10)*, which allow elution at high pH values; CaptAvidin from Molecular Probes, Eugene, OR, USA], or monomeric avidin, which displays a lower binding affinity for biotin (Immobilized Monomeric Avidin, Pierce) *(11)*. Important methodologies in quantitative proteomics, such as Isotope-Coded Affinity Tag (ICAT) technology are based on the use of monomeric avidin *(12)*. However, lowering the affinity of (strept)avidin–biotin interaction prevents an efficient capture of biotinylated molecules in the presence of strong detergents, which are often essential for the solubilization of hydrophobic molecules (e.g., membrane proteins).

We performed a survey of the existing literature for methods allowing quantitative recovery of biotinylated proteins from streptavidin columns. Indeed, some groups have described the elution of biotinylated proteins from (strept)avidin resins using harsh experimental conditions, such as use of guanidinium chloride or elution in sodium dodecyl sulphate (SDS)-containing gel-loading buffer *(13,14)*, though without commenting on yields. Recently, a technical bulletin from Sigma-Aldrich Corporation *(15)* described elution of biotinylated tryptic peptides from streptavidin-coated multiwell plates by means of a solution of 70% acetonitrile, 0.5% formic acid and 1 mM biotin (http://www.sigmaaldrich.com/suite7/Area_of_Interest/Life_Science/Life_Science_Quarterly/Winter_2003.html). This procedure, however, is not in principle readily applicable to proteins.

In this chapter, we present a protocol *(16)* for the quantitative elution of biotinylated proteins from streptavidin sepharose, featuring harsh elution conditions and competition with free biotin. The method is described in detail for the biotinylation of standard purified proteins [taking bovine serum albumin (BSA) as an example], but can easily be extended and applied to the biotinylation of protein extracts, intact cells and even whole tissues and organs. Indeed, the usefulness of the method is exemplified by the recovery of biotinylated proteins from organ homogenates (*see* **Fig. 1**), obtained from mice perfused with a reactive ester derivative of biotin *(17)*.

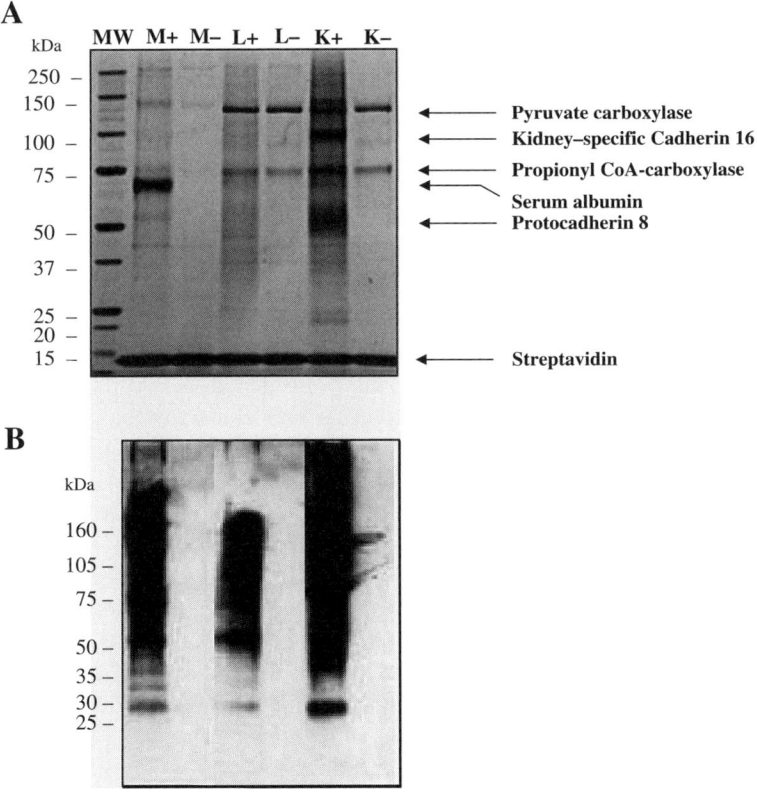

Fig. 1. *(Continued)*

2. Materials

2.1. Biotinylation of Purified Standard Proteins (Bovine Serum Albumin)

1. BSA (MW = 66,000) (cat. no. A 7030, Sigma-Aldrich, St. Louis, MO, USA;) solubilized in phosphate-buffered saline (PBS; NaH_2PO_4 20 mM, Na_2HPO_4 30 mM, NaCl 100mM in MilliQ water, pH 7.4) at the concentration of 2mg/ml (~30 µM).
2. Sulfosuccinimidyl-6-(biotinamido)hexanoate (Sulfo-NHS-LC-biotin, MW = 556.59) (cat. no. 21335, Pierce), dissolved in PBS at 3 mM concentration (a 100:1 molar excess with respect to the BSA concentration).
3. Tris(hydroxymethyl) aminomethane hydrochloride (Tris, MW = 157.6) (cat. no. T 5941, Sigma-Aldrich) dissolved in MilliQ water at 50 mM concentration, pH 7.4.
4. Vivaspin concentrators VS 2002 (MW cut-off 10,000 Da, volume = 20 ml) (Vivascience, Verviers, Belgium).

◀──

Fig. 1. Purification of biotinylated proteins from in vivo perfused mouse organs. The perfusion of anaesthetized mice was performed as described elsewhere *(17)*. Biotinylated proteins were purified from different organs as follows. Organ specimens were homogenized in 20 µl lysis buffer [2% SDS, 50 mM Tris, 10 mM EDTA, complete EDTA-free protease inhibitor cocktail (1 tablet/50 ml, Roche Diagnostics, Mannheim, Germany) in phosphate-buffered saline (PBS) pH 7.4] per mg of tissue using an Ultra-Turrax T8 disperser (IKA-Werke, Staufen, Germany). Homogenates were processed by 5-min sonication (to shear DNA), 5-min incubation at 99°C and 5-min centrifugation at 16,100 g, and the resulting supernatant was taken as total protein extract. Protein concentration was determined as described, and protein extracts were diluted in lysis buffer. Two milligrams of protein (4 mg/ml) were added to 100 µl pre-washed (*see* **Note 4**) streptavidin sepharose resin. Capturing and elution of biotinylated proteins were carried out as described for the bovine serum albumin (BSA) experiments (*see* **Subheadings 3.2.** and **3.3.**). For analysis of the purified proteins, 24 µl of each extract after release from the resin was subjected to sodium dodecyl sulphate–polyacrylamide gel electrophoresis, carried out as described for the BSA experiment (*see* **Subheading 3.4.**). The gel was stained overnight with Sypro Ruby Protein Stain (panel **A**, *see* **Note 6**). Another replicate gel was transferred to nitrocellulose and biotinylated proteins revealed by streptavidin–horseradish peroxidase staining and enhanced chemiluminescent as described in **Subheading 3.5.** (panel **B**). M, skeletal muscle; L, liver; K, kidney; MW, molecular weight markers; +, mice perfused with Sulfo-NHS-LC-biotin; –, control mice perfused with PBS. Some relevant proteins identified by mass spectrometric analysis are indicated on the right. Proteins appearing in control, PBS-perfused mice were identified as enzymes which use biotin as a co-factor and which carry several covalently bound biotin moieties.

2.2. Capture and Release of Biotinylated Proteins from Streptavidin Sepharose

1. Bio-Rad Protein Assay kit (cat. no. 500-0002, Bio-Rad, Hercules, CA, USA).
2. Streptavidin Sepharose™ High Performance (cat. no. 17-5113-01, Amersham Biosciences, Uppsala, Sweden).
3. Buffer A (1% NP-40, 0.1% SDS in PBS).
4. Buffer B (0.1% NP-40, NaCl 0.5 M).
5. Revolving mixer (e.g., Reax-2 from Heidolph Instruments, Schwabach, Germany).
6. Ultrafree-MC centrifugal filters (5.0-μm Durapore filter units) (cat. no. UFC30SV00, Millipore Co., Bedford, MA, USA).
7. Tabletop microcentrifuge (e.g., Eppendorf Mod. 5415 D, Eppendorf Ag, Hamburg, Germany).
8. Elution solution [Urea 6 M, Thiourea 2 M, D-biotin 30 mM (*see* **Note 1**) SDS 2% in PBS, pH 12.0].
9. Heating block (e.g., Thermomixer Comfort, Eppendorf).

2.3. SDS–Polyacrylamide Gel Electrophoresis with Pre-Cast Gels

1. XCell SureLock Mini Cell (cat. no. EI0001, Invitrogen, Carlsbad, CA, USA).
2. NuPAGE Novex Bis-Tris 4–12% acrylamide gradient gels (cat. no. NP0335, Invitrogen). Store at 4°C.
3. MOPS 20× concentrated running buffer (cat. no. NP0001, Invitrogen). Store at room temperature.
4. Reducing loading buffer: 235 mM SDS, 33.6% v/v glycerol, 5% v/v 2-mercaptoethanol, 0.67% w/v bromophenol blue, 210 mM Tris–HCl, pH 6.8. Store at room temperature.
5. Rainbow Molecular Weight Markers (cat. no. RPN800, Amersham Biosciences).

2.4. Western Blotting for Biotinylated Proteins

1. XCell II Blot Module (cat. no. EI9051, Invitrogen).
2. Transfer buffer, (20×): 500 mM Bicine, 500 mM Bis-tris (free acid), 20 mM NaEDTA.
3. Methanol analytical grade.
4. Protran™ nitrocellulose membrane (cat. no. 10 401196, Schleicher and Schuell, Dassel, Germany) and 3MM Chr chromatography paper (cat. no. 3030-861, Whatman, Maidstone, UK).
5. Tris-buffered saline with Tween (TBS-T): Prepare 10× stock with 1.37 M NaCl, 27 mM KCl, 250 mM Tris–HCl, pH 7.4, 1% Tween-20. Dilute 100 ml with 900 ml MilliQ water before use.
6. Blocking buffer: 5% (w/v) nonfat dry milk in TBS-T.
7. Streptavidin conjugated to HRP (cat. no. RPN 1231, Amersham Biosciences).

8. Enhanced chemiluminescent (ECL) reagents from Amersham Biosciences (cat. no. RPN 2132, ECL Plus Western Blotting Detaction System) and Bio-Max Light film (cat. no. 876-1520, Kodak, Rochester, NY, USA).
9. Saran Wrap.
10. X-ray film developing apparatus (e.g., Agfa Curix CP1000 automatic film processor).

3. Methods

3.1. Biotinylation of Bovine Serum Albumin

BSA is dissolved in PBS as described in **Subheading 2.1.**, and 50 µl of BSA solution (~30 µM) is allowed to react for 30 min at room temperature (RT) with an equal volume of a 100-fold molar excess of sulfo-NHS-LC-biotin. Unreacted biotinylation reagent is quenched by adding an excess of a primary amine-containing solution (Tris–Cl 50 mM, pH 7.4, 50 µl). Biotinylated BSA is purified from the reaction by-products by means of repeated washing with PBS using spin dialysis on Vivaspin centrifugal concentrators. Alternatively, biotinylated BSA can be purified by size-exclusion chromatography on Sephadex™ G25 Medium (PD-10 desalting columns, cat. no. 17-0851-01, Amersham Biosciences) (*see* **Note 2**). For monitoring of biotinylation efficiency (*see* **Note 3**).

3.2. Capture of Biotinylated Proteins with Streptavidin Sepharose

The concentration of biotinylated BSA after purification is determined with the Microarray procedure from the BioRad Protein Assay kit, following manufacturer's instructions. Biotinylated BSA is then diluted with PBS to a final protein concentration of 0.2 mg/ml, and an aliquot of 500 µl is added to 72 µl of pre-washed streptavidin sepharose resin (*see* **Note 4**). After 2 h of incubation at RT, the slurry is transferred to an Ultrafree centrifugal filter and spun down for 30 s at 16,100 g in a tabletop centrifuge. The filtrate is removed, and the resin is then washed three times with 400 µl of buffer A, followed by two additional washing steps with 400 µl of buffer B. Finally, the resin is quantitatively removed from the filtering unit with two washing steps (each 50 µl) using PBS and transferred to a fresh tube.

3.3. Quantitative Release of Biotinylated Proteins from Streptavidin Sepharose

The resin is pelleted by a 30-s centrifugation step in the tabletop centrifuge and the supernatant carefully removed. Five hundred microliters of the elution solution is added to the pelleted resin, which is resuspended by mild shaking and incubated for 15 min at room temperature, followed by 15-min incubation in the heating block set at 96 °C, with occasional agitation. The slurry is quantitatively transferred to a new Ultrafree centrifugal filter, spun down for 30 s at 16,100 *g* in the tabletop centrifuge and the filtrate recovered.

3.4. SDS–PAGE of Released Biotinylated Proteins

Extract one NuPAGE Novex Bis-Tris 4–12% acrylamide gel from its package and carefully remove the well-forming comb and the white tape strip at the bottom of the gel. Gently rinse the wells twice with MilliQ water, draining the liquid each time by inversion of the gel. Assemble the gel in the XCell SureLock Mini Cell apparatus, paying attention to correctly seal the inner chamber (*see* **Note 5**). Dilute 30 ml of the 20-fold concentrated MOPS running solution with MilliQ water to a final volume of 600 ml. Fill the inner chamber of the gel with enough buffer as to completely cover the gel wells. Pour the remaining running buffer in the outer chamber. The slot in the bottom part of the gel cassette should be completely submerged in the buffer. The gel is now ready for loading.

Twenty-four microliters of the filtrate (*see* **Subheading 3.3.**) are mixed with 6 µl of loading buffer and heated at 95°C in the heating block for 5 min. After cooling, the sample is applied onto the wells of the SDS–polyacrylamide gel electrophoresis (SDS–PAGE) gel. Ten microliters of the Rainbow Molecular Weight Markers are also loaded in one well. Electrophoretic separation is allowed to proceed at 110 mA, 180 V for 1 h or until the Bromophenol Blue front reaches the bottom of the gel. Gels are then removed from the casting assembly and either processed for Western blotting or fixed for 30 min and stained with Sypro Ruby (*see* **Note 6**).

3.5. Western Blot with Streptavidin–HRP

Prepare 1 l of 1× NuPAGE transfer buffer by adding 50 ml 20× NuPAGE transfer buffer and 200 ml methanol to 750 ml of MilliQ water.

Gels obtained as described in **Subheading 3.4.** are soaked for 10 min at RT in 1× NuPAGE transfer buffer with gentle agitation. During this time, one piece of nitrocellulose membrane and several pieces of 3MM filter paper are cut at 9.5 × 5 cm, the internal dimensions of the inner core of the XCell II Blotting module. The membrane (*see* **Note 7**), the filter paper and the blotting pads (*see* **Note 8**) are also soaked in 1× NuPAGE transfer buffer.

The blotting device is assembled as follows. Two soaked blotting pads are put into the cathode (–) core of the blot module, followed by 4–5 sheets of soaked filter paper and by the soaked gel, flipped by 180° with respect to the original sequence of the lanes (e.g., with respect to the Rainbow markers position). Carefully place the pre-soaked transfer membrane on the gel, remove carefully any trapped air bubbles. Place again 4–5 sheets of soaked filter paper on top and again two soaked blotting pads. Close with the anode (+) core, squeeze firmly together the blot module and slide it into the guide rails of the lower buffer chamber. Fill the blot module with $1\times$ NuPAGE transfer buffer, the outer chamber with water for refrigeration, place the lid on the unit and start the run. Perform the transfer to the nitrocellulose at 30 V constant for 2 h, setting the maximum current at 220 mA.

Once the transfer is complete, the blotting module is disassembled and the membrane is soaked twice at RT in blocking buffer (30 min each) on a rocking platform. Streptavidin conjugated to HRP is diluted 1:1000 in blocking buffer, and 20 ml of this solution is poured onto the membrane. The streptavidin-biotin binding reaction is allowed to proceed at RT for 45 min. The streptavidin–HRP solution is then removed and the membrane washed three times for 10 min each with 50 ml of TBS-T and once with PBS. The remaining steps are performed in a dark room under safe light conditions. The membrane is put of a piece of Saran Wrap, the ECL reagents mixed and poured on the membrane for 1 min. After this time, the excess reagent is drained, the membrane tapped on clean KimWipes and placed between two layers of clean Saran wrap. The membrane is exposed to X-ray film in an X-ray cassette for variable length of time, and films are developed in a developing machine.

4. Notes

1. D-biotin (MW = 244.3; cat. no. B 4639, Sigma-Aldrich) is not readily soluble in aqueous solutions, but can be dissolved at 10 mg/ml in 1 M NaOH, which corresponds roughly to 40 mM. A stock solution (300 mM) of D-biotin for the preparation of elution solution can be prepared by first dissolving 366.5 mg of the vitamin in 400 μl of 8 M NaOH and then diluting this solution with MilliQ water to a final volume of 5 ml.

2. PD-10 desalting columns are pre-packed, disposable columns containing Sephadex™ G25 Medium for group separation of high (MW >5000) from low molecular weight (MW <1000) substances by desalting and buffer exchange. Columns are equilibrated with approximately 25 ml elution buffer (e.g., PBS), prior to addition of the protein sample diluted to a final volume of 2.5 ml with PBS. Elution is performed with 3.5 ml of PBS and the flow-through is collected. Protein yield is typically greater than 95%, with very little residual contamination. Concentration of the flow-through may be carried out by means of Vivaspin concentrators, as described in **Subheading 3.1.**

3. BSA contains 59 lysine residues in its sequence, 52 of which result to be accessible to the biotinylation reagent dissolved in aqueous solution. The level of biotinylation of BSA can be monitored by native gel electrophoresis (PhastGel, Amersham Biosciences) (data not shown). An alternative, more precise way to

evaluate (actually, count) the number of biotinylated lysine residues in a given protein consists in evaluating its mass before and after the biotinylation procedure by means of MALDI-TOF mass spectrometry. **Figure 2** shows one of such measurements, performed at a BSA : biotin ratio of 1:100 with a cleavable

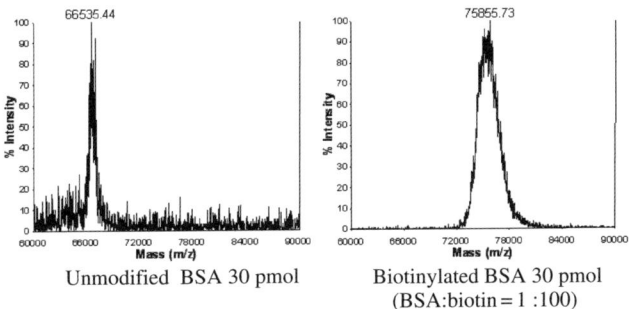

Unmodified BSA 30 pmol

Biotinylated BSA 30 pmol
(BSA:biotin = 1 :100)

[BSA] : [Biotin]	Mass (Da)	Mass difference compared to BSA (Da)*	Number of biotins
1:0	66536.41	0	0
1:1	66948.43	412.02	~1
1:10	68810.27	2273.86	~6
1:100	75855.73	9319.32	~24

BSA: total 59 lysines , ~52 are accessible from the aqueous solution.
* One SS-Biotin residue has a mass of 390.10 Da.

Fig. 2. Determination of protein biotinylation by MALDI-TOF analysis. Bovine serum albumin (BSA) was biotinylated as described in **Subheading 3.1.**, only sulfos-uccinimidyl 2-(biotinamido)-ethyl-1, 3-dithiopropionate (Sulfo-NHS-SS-biotin, MW = 606.69) (cat. no. 21331, Pierce) was used instead of sulfo-NHS-LC-biotin. One micro-liter of unmodified BSA or of biotinylated BSA was mixed with 1 µl of a solution of MALDI matrix (3 mg/ml α-cyano 4-hydroxy cinnamic acid in 0.03% trifluoroacetic acid (TFA) and 70% acetonitrile), and 1 µl of this mixture was deposed on a 192-well MALDI plate. MALDI-TOF analysis was carried out with the 4700 Proteomics Analyzer (Applied Biosystems, Framingham, MA, USA). Spectra were acquired with a Nd:YAG laser working at a laser frequency of 200 Hz. The table in the lower part of the figure summarizes the mass shifts observed performing the same experiment with different BSA : biotin ratios and the average number of modified lysine residues in BSA at each ratio.

linker-containing form of the biotinylation reagent (Sulfo-NHS-SS-biotin). The mass shift of 9319.32 Da accounts for about 24 biotin residues bound to lysines in the protein.

4. Streptavidin Sepharose High Performance slurries contain ethanol as a preservative, which has to be removed completely before use. The desired amount of resin (1 ml of slurry binds 300 nmol of biotin) is therefore pre-washed twice with wash buffer A (1% NP-40, 0.1% SDS in PBS) and twice with PBS. After each incubation or washing step, the sepharose is pelleted by short centrifugation in a tabletop microcentrifuge. The pellet being extremely loose, attention should be paid not to aspirate Sepharose particles together with the liquid when removing the supernatant.

5. The XCell SureLock Mini Cell apparatus can accommodate two NuPAGE gels at a time. Attention should be paid to correctly orient the gel, in order to have the U-shaped plastic side of each gel cassette placed toward the inner core of the apparatus. If only one gel is needed, a plastic part of the same shape of a gel cassette (provided) can be used to close the opposite side of the inner core and form the inner buffer chamber.

6. For sensitive and efficient staining of gels, Sypro Ruby represents a very useful, though expensive, alternative to Coomassie Blue or Silver staining procedures. Once removed from the gel cassette, gels are soaked in fix solution (10% methanol, 7% acetic acid) for 30 min at RT and then stained for 2 h (or overnight) with Sypro Ruby Protein Stain (cat. no. S12000, Molecular Probes/Invitrogen). After staining, the Sypro Ruby solution is recovered and can be reused for at least two times, without significant loss of activity. The gels are destained by repeated changes of fix solution (30 min each) on a shaker and finally rinsed in MilliQ water. After destaining, gels can be imaged using a UV Transilluminator and a suitable recording device (e.g., DIANA 2 imager, Raytest, Straubenhardt, Germany).

7. Nitrocellulose membranes should be carefully pre-wetted, in order to avoid air micro-bubble trapping in the nitrocellulose, with consequent irregular buffer flow during transfer. The nitrocellulose membrane pieces should be gently laid on the surface of the transfer buffer liquid, instead of submerging them in liquid immediately or pouring liquid on top of the membrane itself. Allow the liquid to be absorbed by the membrane from the bottom side for a few minutes, until tiny droplets appear at the upper surface. The membrane can then be soaked in the liquid bulk without further problems.

8. Air bubble trapping can represent a major problem for a correct transfer of proteins from gels to blotting membranes. Very often, blotting pads are a major source of air bubble trapping, especially when they have already been used several times, and care should be taken to eliminate air from them completely. In our experience, wetting blotting pads with deionized water under pressure helps dislodging also very tiny air bubbles. Blotting pads can then be gently squeezed to eliminate as much water as possible and soaked in 2–3 changes of 1× NuPAGE transfer buffer.

Acknowledgments

Financial support of the Gebert-Rüf Foundation, the Swiss National Science Foundation and of the Bundesamt für Bildung und Wissenschaft (EU Projects Angiogenesis and Stroma FPS/6 no. 503233) and the access to instrumentation of the Functional Genomics Center Zurich are gratefully acknowledged. Giuliano Elia is on leave of absence from Institute of Neurobiology and Molecular Medicine CNR, Rome, Italy.

References

1. Boas, M. A. (1927) The effect of desiccation upon the nutritive properties of egg-white. *Biochem. J.* **21**, 712–724.
2. György, P., Rose, C. S., Hofmann, K., Melville, D. B., Du Vigneaud, V. (1940) A further note on the identity of vitamin H with biotin. *Science* **92**, 609.
3. Du Vigneaud, V., Melville, D. B., György, P., Rose, C. S. (1940) On the identity of vitamin H with biotin. *Science* **92**, 62–63.
4. György, P., Rose, C. S., Eakin, R. E., Snell, E. E., Williams, R. J. (1941) Egg-white injury as the result of nonabsorption or inactivation of biotin. *Science* **93**, 477–478.
5. Pennington, P., Snell, E. E., Eakin, R. E. (1942) Crystalline avidin. *J. Am. Chem. Soc.* **64**, 469.
6. Wilchek, M. and Bayer, E. A. (1990) Introduction to avidin-biotin technology. *Methods Enzymol.* **184**, 5–13.
7. Wilchek, M. and Bayer, E. A. (1990) Applications of avidin-biotin technology: literature survey. *Methods Enzymol.* **184**, 14–45.
8. Green, N. M. (1975) Avidin. *Adv. Protein. Chem.* **29**, 85–133.
9. Olejnik, J., Sonar, S., Krzymanska-Olejnik, E., Rothschild, K. J. (1995) Photo-cleavable biotin derivatives: a versatile approach for the isolation of biomolecules. *Proc. Natl. Acad. Sci. U. S. A.* **92**, 7590–7594.
10. Morag, E., Bayer, E. A., Wilchek, M. (1996) Immobilized nitro-avidin and nitro-streptavidin as reusable affinity matrices for application in avidin-biotin technology. *Anal. Biochem.* **243**, 257–263.
11. Ellerbroek, S. M., Wu, Y. I., Overall, C. M., Stack, M. S. (2001) Functional interplay between type I collagen and cell surface matrix metalloproteinase activity. *J. Biol. Chem.* **276**, 24833–24842.
12. Gygi, S. P., et al. (1999) Quantitative analysis of complex protein mixtures using isotope-coded affinity tags. *Nat. Biotechnol.* **17**, 994–999.
13. Brandli, A. W., Parton, R. G., Simons, K. (1990) Transcytosis in MDCK cells: identification of glycoproteins transported bidirectionally between both plasma membrane domains. *J. Cell. Biol.* **111**, 2909–2921.
14. De La Fuente, E. K., et al. (1997) Biotinylation of membrane proteins accessible via the pulmonary circulation in normal and hyperoxic rats. *Am. J. Physiol.* **272**, L461–L470.

15. Dapron, J., Zobrist, J., Foster, K., Barbacci, L., Hassell, T., Kappel, W. (2003) Proteomic applications of specialty high capacity streptavidin-coated multiwell plates. *Life Science Quarterly*, **4**, http://www.sigmaaldrich.com/img/assets/10120/Protec_-_Streptavidin.pdf.
16. Rybak, J. N., Scheurer, S. B., Neri, D., Elia, G. (2004) Purification of biotinylated proteins on streptavidin resin: a protocol for quantitative elution. *Proteomics* **4**, 2296–2299.
17. Rybak, J.-N., et al. (2005) In vivo protein biotinylation for the identification of organ-specific antigens accessible from the vasculature. *Nat. Methods* **2**, 291–298.

9

High-Yield Production and Purification of Recombinant T7-Tag Mature Streptavidin in Glucose-Stressed *E. coli*

Nicolas Humbert, Peter Schürmann, Andrea Zocchi, Jean-Marc Neuhaus, and Thomas R. Ward

Summary

The overexpression of toxic recombinant proteins is often problematic, leading to either low production levels or inclusion bodies. Streptavidin is no exception and thus the highest production level reported to date for streptavidin is 70 mg/L of functional protein. Herein, we report on the production in *Escherichia coli* and the purification of a recombinant mature streptavidin bearing a T7-tag. Optimization of critical parameters, including the glucose concentration, the pH and the time of induction as well as the use of BL21(DE3)pLysS cell strain, affords up to 120 mg/L functional streptavidin in soluble form. The yield can be further increased by an osmotic stress during the preculture by adding highly concentrated glucose before the inoculation of the culture medium, thus affording reproducibly 230 mg/L of soluble streptavidin. A single denaturing-renaturing step and affinity chromatography afford highly active tetrameric protein with >3.8/4.0 active sites.

Key Words: Streptavidin; *E. coli*; affinity purification; biotin; osmotic stress.

1. Introduction

The production of toxic recombinant proteins by *Escherichia coli* can be challenging due to either the poisoning of the cells, which stop growing, or the formation of inclusion bodies which may be difficult to renature *(1)*. In the case of recombinant streptavidin, the cell growth is inhibited by the deprivation of free biotin, an essential vitamin for *E. coli*, as well as by the presence of bacterial biotinylated proteins *(2,3)*.

From: *Methods in Molecular Biology, vol. 418: Avidin-Biotin Interactions, Methods and Applications*
Edited by: R. J. McMahon © Humana Press, Totowa, NJ

Our research on artificial metalloenzymes based on the biotin–streptavidin technology requires large amounts of streptavidin for screening purposes *(4–8)*. Due to the prohibitive price of this protein (US$100 per mg), it was indispensable for us to improve on the published production levels.

Previous reports on streptavidin production in different microorganisms including *E. coli* have been published over the years *(9–13)*. To the best of our knowledge, the highest yield obtained with *E. coli* was 60–70 mg/L of culture of pure protein, using BL21(DE3)–pET11b system *(11)*. In our hands, the protein expression in this strain using the plasmid pET11b-SAV *(11)* encoding for the T7-Tag mature streptavidin (*see* **Fig. 1**) is poorly regulated, and depending on the medium's provider, the presence of the protein can be observed at an early state of the bacterial development (i.e., before the cells are inducible) suggesting an activation of the *lac* operon promoter. This premature protein expression leads to cell death and thus low production yields (0–8 mg/L of culture, unpublished results).

For this reason, we selected the BL21(DE3)pLysS strain, supplemented by an optimized glucose concentration *(10)*. Using a fermentor, this system allowed us to reach 60–70 mg/L functional streptavidin. After a broad screen of culture media, we observed that the quantity of protein per bacterium is very similar under all the conditions (data not shown). This suggests that an optimal growth of bacteria is required to obtain a maximal protein production.

The best conditions are cultivation of the cells in Modified Tryptone-Phosphate Medium (MTP) (*see* **Note 6**) in the presence of 0.4% glucose at 37°C. The cells are induced at $OD_{600\,nm}$ = 1.800–2.200 (~3 h after the start of the culture) and the culture is stopped 3 h after induction. With these conditions, we routinely obtain a yield of 100–120 mg/L of culture of a highly pure and approximately 100% active protein.

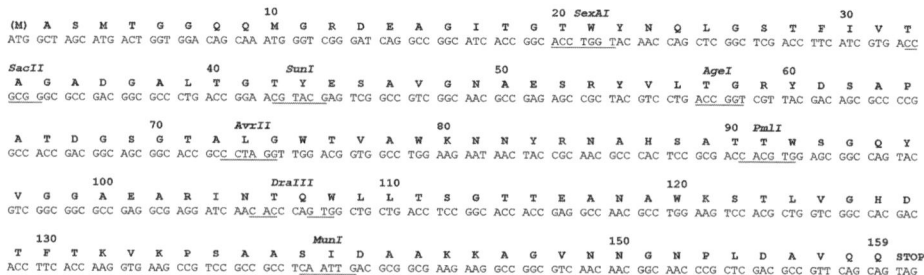

Fig. 1. Redesigned T7-Tag mature streptavidin gene sequence, including the unique restriction sites (underlined) and the corresponding aminoacid sequence. Upon expression, the initial methionine (in parentheses) is cleaved. The T7-Tag (aminoacid residues 1–12) is italicized.

By mixing together the preculture with a high concentration of glucose (i.e., the total quantity of glucose required for the culture), before inoculation of the medium, the yield can be reproducibly increased to 210–230 mg/L of culture. In the present work, we present an optimized production protocol.

2. Materials
2.1. Cell Culture and Lysis

1. Stock solution of glucose (20%, 180.2 g/mol): D-glucose (200 g) is dissolved in distilled water (final volume = 1 L). The solution is aliquoted in 200 mL fractions in autoclavable bottles and then autoclaved (20 min, 120°C, 1.5 bar, *see* **Note 1**). This solution can be stored at room temperature for a few months.
2. Ampicillin stock solution (50 mg/mL): Ampicillin sodium salt is diluted in nanopure water, is filtered (0.22-µm filters), aliquoted in 12 mL fractions and stored at –20°C for a few months.
3. Chloramphenicol stock solution (34 mg/mL): Chloramphenicol is diluted in 70% ethanol solution, filtered (0.22-µm filters), aliquoted in 10 mL fractions and stored at –20°C for a few months.
4. Isopropyl-β-D-thiogalactopyranoside (IPTG) stock solution: IPTG powder (MW = 238.3 g/mol) is weighed and dissolved in nanopure water to a final concentration of 0.8 M. The solution is then filtered (0.22-µm filters), aliquoted in 5 mL fractions and stored at –20°C for a few months.
5. LB, Ampicillin, Chloramphenicol, Glucose (LACG) Petri dishes: The LB powder (Brunschwig) is mixed together with bacto-agar (Difco) and dissolved with approximately 450 mL of distilled water in a 500-mL autoclavable bottle. After autoclaving (*see* **Note 2**, 20 min, 120°C, 1.5 bar) and cooling to 50–60°C, the glucose stock solution (25 mL), the ampicillin stock solution (600 µL) and the chloramphenicol stock solution (500 µL) are added sequentially (final concentrations: 1%, 60 µg/mL and 34 µg/mL, respectively). The volume is adjusted with sterile nanopure water to 500 mL. After mixing, the medium is poured on Petri dishes (10–15 mL each). Once the medium is solidified, the dishes are stored at 4°C for up to 4 months (*see* **Note 3**).
6. MTP medium for a 500-mL preculture: Bactotryptone (10 g), Na_2HPO_4 (1.11 g), KH_2PO_4 (500 mg), NaCl (4 g) and bacto-yeast extract (7.5 g) are dissolved in distilled water (500 mL), autoclaved (20 min, 120°C, 1.5 bar) and stored at room temperature for up to 1 year (*see* **Note 4**).
7. MTP medium for a 10 L fermentation: Bactotryptone (200 g), Na_2HPO_4 (13 g), KH_2PO_4 (10 g), NaCl (80 g) and bacto-yeast extract (150 g) are mixed together, dissolved in distilled water, autoclaved (20 min, 120°C, 1.5 bar) and stored at room temperature for up to 1 year.
8. Antifoam A is provided by Sigma.
9. Resuspension buffer: 20 mM Tris–HCl, pH 7.4, sodium azide [0.02% (w/v)] and $MgCl_2$ (10 mM). The solution is stored at room temperature for up to 1 year.

10. DNaseI solution (1 mg/mL): Dissolve 10 mg of DNaseI in storage buffer (5 mM Tris–HCl, pH7.5, 75 mM NaCl, 0.5 M MgCl$_2$, 50% glycerol). Aliquot in 1 mL fractions and store at –20°C.

2.2. SDS–PAGE

1. Streptavidin stock solution (152 µM = 10 mg/mL, MW 65 700 g/mol). The pure protein (10.0 mg) is weighed in a 1.5-mL microcentrifuge tube and dissolved in nanopure water (1 mL). The solution is aliquoted (50 µL in 500-µL microcentrifuge tubes) and stored for up to 1 year at –20°C.
2. Separating buffer (4×): 1.5 M Tris–HCl, pH 8.8. Store at 4°C.
3. Stacking buffer (4×): 0.5 M Tris–HCl, pH 6.8. Store at 4°C.
4. Acrylamide/bisacrylamide solution (37.5:1) was purchased from National Diagnostics. It should be manipulated with a great care due to its high toxicity (neurotoxic and carcinogenic). It should be stored in a well-ventilated hood at room temperature.
5. SDS solution, 10%: as SDS is highly volatile and irritant, it is strongly advised to weigh it wearing glasses, gloves and a mask. Dissolve it in distilled water and store at room temperature.
6. N,N,N´,N´-Tetramethyl-ethylenediamine (TEMED). Store at 4°C.
7. Ammonium persulfate: prepare a 10% (w/v) solution in distilled water, store at 4°C for immediate use or at –20°C for long-term storage (*see* **Note 5**).
8. Denaturing gel loading buffer (5×): 0.25 M Tris–HCl, pH 6.8, 0.03% (w/v) Bromophenol Blue, 50% sucrose, 2% v/v β-mercaptoethanol, 1% w/v SDS. Store at –20°C.
9. Running gel buffer (10×): 0.25 M Tris, 1.92 M glycine, 1% (w/v) SDS. Store at room temperature.
10. Gel fixing solution: 25% v/v isopropanol; 10% acetic acid. Store at room temperature in a well-ventilated hood.
11. Gel staining solution: 0.25% w/v coomassie brilliant blue; 45% v/v methanol; 10% v/v acetic acid. Store at room temperature in a well-ventilated hood.
12. Gel destaining solution: 5% v/v methanol; 7% v/v acetic acid. Store at room temperature in a well-ventilated hood.

2.3. Treatment of the Bacterial Extracts

1. Denaturing solution: 6 M guanidium hydrochloride, pH 1.5.
2. Renaturing solution: 20 mM Tris–HCl, pH 7.4.
3. Iminobiotin-binding buffer: 50 mM Na$_2$CO$_3$, pH 9.8, 0.5 M NaCl (*see* **Note 6**).
4. Elution solution: 0.1 M acetic acid, pH 2.9 (approximately 1% acetic acid).
5. Iminobiotin storage buffer: 10 mM sodium phosphate, pH 6.8, 0.5 M NaCl, 0.02% sodium azide.
6. Neutralizing buffer 1: 2 M Tris–HCl, pH 7.4.
7. Neutralizing buffer 2: 10 mM Tris–HCl, pH 7.4 (*see* **Note 7**).

3. Methods

As the streptavidin is a very toxic protein, its production by *E. coli* should be strictly regulated. The protein expression from the pET11b plasmid is dependent on the presence of T7 polymerase, which is itself controlled by the promotor of the *lac* operon. The combination of BL21(DE3)pLysS cells and an optimized glucose concentration allows total suppression of the streptavidin expression before induction (*see* **Fig. 2**).

In this procedure, particular care is required concerning the sterility during the plating of the cells, the preculture and the inoculation of the medium. During the purification of the protein, we strongly advise to keep the sample at 4°C as much as possible after the renaturation step.

3.1. Cell Cultures

1. Freshly transformed BL21(DE3)pLysS cells with either a wild-type or a mutated plasmid containing the streptavidin gene or glycerol stock are plated on LACG dishes and incubated overnight (12–16 h) at 37°C. The dishes are sealed with a parafilm strip and stored at 4°C for up to 1 week.
2. One single circular colony (*see* **Note 8**) is picked and used to inoculate the preculture medium (150 mL) in 1-L Erlenmeyer flask (*see* **Note 9**).
3. Incubate overnight (12–16 h) the preculture at 37°C, under orbital shaking (250 rpm). To avoid contamination by bacteria having lost their plasmid, the preculture should be used immediately after incubation.

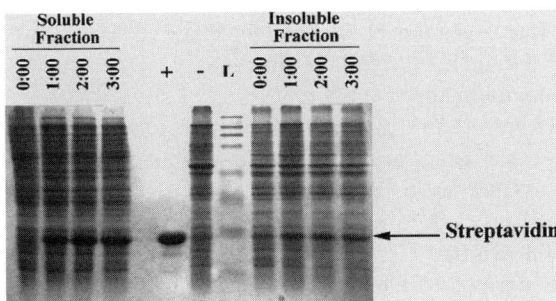

Fig. 2. Denaturing SDS–PAGE analysis of a 10 L fermentation of *Escherichia coli* expressing T7-Tag mature streptavidin. Soluble and insoluble fractions are loaded separately and the times indicated start from the moment of induction. + indicates the purified streptavidin, – indicates bacterial extract of *E. coli* which do not express the streptavidin and L is the Broad Range Protein Marker (Bio-Rad).

4. Before inoculation, add ampicillin and chloramphenicol to the medium (using the pre-mode aliquots, to a final concentration of 60 μg/ml and 34 μg/ml, respectively), mix-in 10 mL of antifoam A (*see* **Note 10**) and heat gently to a temperature of 37°C

5. Mix the preculture with the entire glucose solution (*see* **Note 11**) and inoculate the medium.

6. Stir the culture at 600 rpm and set the aeration to 20 L/min (*see* **Note 12**).

7. To follow the bacterial growth, collect a few milliliters of the culture and measure the OD at 600 nm every hour (*see* **Note 13**). The growth should be very slow at the beginning and become exponential after 90–120 min.

8. When the $OD_{600\,nm}$ reaches 1.8–2.2, induce the bacteria with an aliquot of IPTG and collect 1 mL of the culture for Sodium Dodecyl Sulphate-Poly Acrylamide Gel Electrophoresis (SDS-PAGE) analysis.

9. Collect 1 mL of culture every hour for a SDS–PAGE analysis. Centrifuge the sample immediately (5 min,13,000 g,room temperature). Discard the medium and store the pellet at –20°C until use.

10. Stop the culture after 3 h of induction (*see* **Note 14**).

11. Centrifuge for 5 min, 5000 g, at 4°C and remove the supernatant (*see* **Note 15**).

12. Take a small aliquot of buffer (10–50 ml) and resuspend gently the cells; add once again a small amount of buffer and gently mix until all cells are resuspended.

13. Pool all the cells in a single centrifugation bottle, adjust the volume of the bacterial suspension to 200 mL with the resuspending buffer and centrifuge (15 min, 5000 g, 4°C).

14. Remove the supernatant and store at –20°C until use.

3.2. Treatment of the Samples Before SDS–PAGE

1. Thaw the samples for analysis and add the resuspending solution. The required volume of the resuspending solution (in μL) is determined by multiplying the $OD_{600\,nm}$ by a factor 40.

2. Add the DNaseI solution (1 μL) and incubate at room temperature until total digestion of the nucleic acids (*see* **Note 16**).

3. Vortex at maximum speed for 5 min.

4. Centrifuge (5 min,16,000 g, room temperature) to pellet the cell debris.

5. Transfer the supernatant in a new centrifuge tube and resuspend the cells debris in the same volume as in **step 1**.

6. Transfer 12 μL of each sample in a new centrifuge tube and add the denaturing loading buffer (3 μL).

7. As positive control, dilute the stock solution of streptavidin (1.2 μL) in nanopure water (10.8 μL). As a negative control, use a bacterial extract of BL21(DE3)pLysS cells which do not express the streptavidin. Add 3 μL of denaturing buffer.

8. Boil all the tubes and the protein ladder at 95°C for 20 min (*see* **Note 17**).

9. Place the tube in ice (*see* **Note 18**).

10. Spin down the liquid by a short centrifugation (15 s).

3.3. SDS–PAGE

1. The following instructions assume the use of the Mini PROTEAN 3 gel system (Bio-Rad), but are valid with other systems that allow the loading of 15-µL samples. Clean the glass plates (2–3 times) with a moist tissue (distilled water) to remove polyacrylamide residues of previous experiments, rinse with ethanol and air-dry.

2. Prepare a 1.5-mm-thick 16% running gel by mixing the 4× separating buffer (2.5 mL) with the acrylamide/bisacrylamide solution (5.3 mL), 10% SDS solution (100 µL) in distilled water (2.05 mL). Following this, add 10% ammonium persulfate (34 µL) and TEMED (5µL) and mix the solution before pouring the gel immediately up to 2–2.5 cm under the higher border, leaving space for the stacking gel *(14)*. Overlay the gel mixture with distilled water or isopropanol. The polymerization of the gel should be complete within 40 min at room temperature.

3. Remove the supernatant liquid, rinse the top of the gel twice with distilled water.

4. Prepare the stacking gel by mixing the 4× stacking buffer (1.5 mL) with the acrylamide/bisacrylamide solution (1 mL), the 10% SDS (60 µL) and water (3.4 mL). As above, the 10% ammonium persulfate (60 µL) and the TEMED (9 µL) should be added last. Mix the solution, pour the gel immediately and insert the comb gently.

5. Prepare the 1× gel running buffer (1 L) by diluting 10 times the 10× stock solution in a 1-L bottle. Mix by inverting the bottle 2–3 times.

6. Once the stacking gel is polymerized, remove carefully the comb and rinse twice the wells with 1× gel running buffer using a 1-mL pipette.

7. After assembling the gel holder, ensure the water tightness, place the assembly in the chamber and fill the chamber with the 1× gel running buffer.

8. Load the samples (including the protein marker) with a micropipette.

9. Complete the assembly and connect the gel unit to a power supply. Run the gel at constant voltage (200 V) and stop when the loading buffer dye reaches the bottom of the gel (45–60 min).

10. Disassemble the setup and put the gel in a small, flat trough containing the fixing solution.

11. Place the container in a microwave oven and heat the gel for 5 min at medium power.

12. Transfer the gel to the staining solution and heat it in the microwave oven for 5 min at medium power (the solution can be used for more than 1 year).

13. Transfer the gel to the destaining solution and heat it in the microwave oven for 20 min at medium power, replace the destaining solution at least once, heat in the microwave oven for 20 min at medium power.

14. Replace one more time the destaining solution and incubate overnight at room temperature until total disappearance of free Coomassie Blue dye. **Figure 2** displays an example of a typical SDS–PAGE for a 10 L fermentation.

3.4. Purification of the Protein

1. Thaw the bacterial pellet and add the minimal volume of resuspending buffer required to cover the sample completely and resuspend the cell lysate.

2. Add 2–3 mg of DNAseI powder (approximately 4000–6000 Units) and incubate at room temperature under gentle stirring until the nucleic acids are totally degraded (*see* **Note 16**).
3. Prepare a large dialysis bag capable of containing at least 400 mL of liquid.
4. Pour the bacterial extract into the dialysis bag. To recover the maximum amount of protein, rinse the vessel that contained the extract with a filtered denaturing solution (guanidium hydrochloride) and also pool this wash-out with the extract into the dialysis bag, before filling to 400 mL (*see* **Note 19**).
5. Place the bag in a tank containing the denaturing buffer (15 L, *see* **Note 20**) and dialyze under gentle stirring (24 h, at room temperature).
6. Transfer the dialysis bag in the renaturing buffer (15 L) and dialyze under gentle stirring (24 h, at room temperature) (*see* **Note 21**).
7. Transfer the dialysis bag to an iminobiotin-binding buffer (15 L) and dialyze under gentle stirring (24 h, 4°C).
8. Centrifuge the extract (4°C, 47000 g, 30 min) and discard the cell debris.
9. Filter the supernatant (0.45 μm).
10. Equilibrate the iminobiotin column with the iminobiotin-binding buffer (5 times the volume of the matrix).
11. Apply the sample to the column.
12. Wash the column with the iminobiotin-binding buffer until the $OD_{280\,nm} = 0$.
13. Elute the protein with the elution buffer and collect the pure protein in 50-mL falcon tubes containing the neutralization solution 1 (2.5 mL; 2 M Tris–HCl, pH7.4).
14. Apply 5 column volumes of elution buffer and equilibrate the column with the iminobiotin storage buffer.
15. Dialyze the pure protein under gentle stirring against the neutralization solution 2 (8 L, 10 mM Tris–HCl, pH7.4) (24 h, 4°C).
16. Transfer the dialysis bag in a tank containing distilled water (8 L) and dialyze under gentle stirring (24 h, 4°C).
17. Transfer the bag in a tank containing nanopure water (5 L) and dialyze under gentle stirring (24 h, 4°C).
18. Filter the protein (0.22 μm) and freeze the solution at –80°C (*see* **Note 22**).
19. Lyophilize the protein (*see* **Note 23**).
20. Store at 4°C for up to few months (*see* **Note 24**).

4. Notes

1. All the stock solutions are aliquoted for 10 L fermentations and the volumes can be readily scaled for different culture volumes.
2. After autoclaving, if antibiotics and glucose have not been added, the medium can be stored as a solid at room temperature in a bottle; it can be melted by microwave heating and then treated as described in **Subheading 2**. Another alternative consists in storing the medium at 55–60°C (agar not solidified). In this case, the glucose and the antibiotic can be directly added just before pouring the liquid in the Petri dishes.

3. To ensure the sterility of the Petri dishes, incubate one of them at 37°C overnight. If one or more bacterial colonies appear, the batch of dishes is probably contaminated and should be discarded.

4. MTP corresponds to the published TP *(11)* but with a slightly modified pH. The pH of TP is about 7.5, as pH of MTP is around 6.9.

5. The ammonium persulfate solution is usually stored at –20°C, but if it is frequently used, it can be stored at 4°C for 2–3 weeks. The solution should be discarded if the polymerization of the gel takes longer than 1 h.

6. This solution can be stored as a 5× concentration.

7. This solution can be stored as a 20× concentration.

8. Ideally, the picked colony should be as circular as possible (which is not always the case because BL21(DE3)pLysS cells tend to form irregular colonies). Concerning the colony size, it should be 1–3 mm in diameter. If they are too big, a large amount of plasmid-lost bacteria may be present; if they are too small, they may be satellite colonies.

9. It is preferable to use a baffled Erlenmeyer flask with a large opening to have a good aeration for the cells.

10. The culture tends to foam abundantly after the induction. Check the presence of foam regularly (at least twice an hour) and do not hesitate to add antifoam when required. For a 10-L culture, it is sometimes required to use 40–50 mL antifoam.

11. During this operation, the concentration of glucose is approximately 12% which we suspect causes an osmotic stress and a total inhibition of the protein expression. In the case of streptavidin, this osmotic stress has doubled the production yield.

12. If not working with a fermentor, use preferentially a baffled Erlenmeyer flask with a large opening and stir on an orbital shaker at 200–250 rpm.

13. The Beer–Lambert law is valid only up to OD < 0.35. Passed this value, it is recommended to dilute the sample with the medium to determine the bacterial turbidity.

14. It is detrimental to pursue the culture after 3 h of induction. In our hands, after 3.5 h, the total amount of cells as well as the isolated protein decreases dramatically. At the end of the culture, the $OD_{600\,nm}$ may reach >15.

15. As it can be difficult to centrifuge 10 L at once, store the cells to be centrifuged at 4°C.

16. Nucleic acids make the extract viscous, their digestion can be considered total, when the bacterial extract becomes fluid.

17. Due to streptavidin's high stability and the presence of endogenous biotin, at least 20 min of thermal denaturation in the presence of SDS is required.

18. Ensure that the tubes are tightly sealed, as they sometimes open in ice and the liquid may escape.

19. The volume of the denaturation dialysis should be adapted to the amount of protein. In practice, the denaturation may be incomplete if the protein's concentration exceeds 5 mg/mL within the dialysis bag *(15)*.

20. A high purity of guanidium hydrochloride is not required (98% purity suffices). Nevertheless, the solution used to rinse the vessel and to complete the volume of sample should be filtered on a buchner funnel under vacuum to remove the insoluble impurities. Moreover, the dialysis solution can be reused 3–4 times.
21. Contrary to the denaturing solution, we advise to use the renaturing solution only once because of contamination of the solution by microorganisms.
22. In rare cases, a slight turbidity appears upon dissolution of the protein in water. We suspect that the precipitate is non-functional, aggregated protein which can readily be removed by filtration if required.
23. As the lyophilized protein is very light and fluffy, caution is required when manipulating the streptavidin powder.
24. We have never observed any significant loss of activity upon prolonged storing (up to 1 year) at 4°C. The streptavidin activity is best assessed using the biotin-4-fluorescein titration published by Gruber *(16)*.

References

1. Novagen (2003) pET System Manual, http://www.novagen.com or http://www.merckbiosciences.co.uk/docs/docs/PROT/TB055.pdf.
2. Green, N. M. (1990) *Methods Enzymol.* **184,** 51–67.
3. Tausig, F., and Wolf, F. J. (1964) *Biochem. Biophys. Res. Commun* **14,** 205–9.
4. Klein, G., Humbert, N., Gradinaru, J., Ivanova, A., Gilardoni, F., Rusbandi, U. E., and Ward, T. R. (2005) *Angew. Chem. Int. Ed. Engl.* **44** (47), 7764–7767.
5. Skander, M., Humbert, N., Collot, J., Gradinaru, J., Klein, G., Loosli, A., Sauser, J., Zocchi, A., Gilardoni, F., and Ward, T. R. (2004) *J. Am. Chem. Soc.* **126,** 14411–8.
6. Letondor, C., Humbert, N., and Ward, T. R. (2005) *Proc. Natl. Acad. Sci. U. S. A.* **102,** 4683–87.
7. Collot, J., Gradinaru, J., Humbert, N., Skander, M., Zocchi, A., and Ward, T. R. (2003) *J. Am. Chem. Soc.* **125,** 9030–1.
8. Collot, J., Humbert, N., Skander, M., Klein, G., and Ward, T. R. (2004) *J. Organomet. Chem.* **689,** 4868–71.
9. Suter, M., Cazin, J., Jr., Butler, J. E., and Mock, D. M. (1988) *J. Immunol. Methods* **113,** 83–91.
10. Sano, T., and Cantor, C. R. (1990) *Proc. Natl. Acad. Sci. U. S. A.* **87,** 142–6.
11. Gallizia, A., de Lalla, C., Nardone, E., Santambrogio, P., Brandazza, A., Sidoli, A., and Arosio, P. (1998) *Protein Expr. Purif.* **14,** 192–6.
12. Bayer, E. A., Ben-Hur, H., Gitlin, G., and Wilchek, M. (1986) *J. Biochem. Biophys. Methods* **13,** 103–12.
13. Nagarajan, V., Ramaley, R., Albertson, H., and Chen, M. (1993) *Appl. Environ. Microbiol* **59,** 3894–8.
14. Laemmli, U. K. (1970) *Nature* **227,** 680–5.
15. Chilkoti, A., Tan, P. H., and Stayton, P. S. (1995) *Proc. Natl. Acad. Sci. U. S. A.* **92,** 1754–8.
16. Kada, G., Falk, H., and Gruber, H. J. (1999) *Biochim. Biophys. Acta* **1427,** 33–43.

10

Endogenous Biotin in Rat Brain
Implications for False-Positive Results With Avidin–Biotin and Streptavidin–Biotin Techniques

Bruce E. McKay, Michael L. Molineux, and Ray W. Turner

Summary

The interaction between avidin and biotin or streptavidin and biotin forms the basis of several widely used immunohistochemical techniques. An assumption inherent to these techniques is that endogenous biotin is not present in the tissue in detectable quantities, as neither avidin nor streptavidin can discriminate between endogenous biotin and biotinylated antibodies. However, biotin is a known cofactor for numerous carboxylases required in oxidative metabolism, raising the possibility for potential false-positive results in many tissues. This issue has been appreciated in liver and kidney tissue, but has received very little attention in nervous tissue. To address this concern, we examined the distribution of biotin throughout the rat central nervous system using avidin- and streptavidin-based detection systems, as well as a monoclonal antibody raised against biotin. Significant levels of endogenous biotin were identified within specific neuronal types, particularly in neurons associated with the cerebellar motor system and the brainstem auditory system. Non-specific (non-biotin) interactions of avidin and streptavidin conjugates with rat brain tissue were further identified and were most pronounced in the lower brainstem. The binding of avidin- and streptavidin-conjugated markers to endogenous biotin and other non-specific interactions with neural tissue were overcome by several methods including the use of blocking kits, prolonged post-fixation of the tissue in paraformaldehyde, or omission of Triton X-100 from the working solution. Without these measures, a reliable estimate of immunolabel may only be achieved in certain brain regions with markers conjugated directly to secondary antibodies.

Key Words: Avidin; streptavidin; endogenous biotin; blocking; immunohistochemistry; cerebellum; auditory system.

From: *Methods in Molecular Biology, vol. 418: Avidin-Biotin Interactions, Methods and Applications*
Edited by: R. J. McMahon © Humana Press, Totowa, NJ

1. Introduction

Avidin–biotin or streptavidin–biotin immunohistochemistry is widely used to localize tissue antigens, study cell morphology, and perform protein analyses. The technique is based on the extremely high affinity binding between the egg white protein avidin or the bacterial protein streptavidin and the vitamin molecule biotin *(1–10)*. In a typical immunohistochemical reaction series, a biotinylated agent (either a primary or a secondary antibody) is detected within tissue by a marker conjugated to avidin or streptavidin. The interaction between (strept)avidin and biotin is extremely stable *(11,12)*. (Strept)avidin–biotin immunohistochemistry has achieved widespread use because biotin or (strept)avidin can be conjugated to a variety of host molecules (e.g., fluorochromes, antibodies, and so on) without changing the properties of the host molecule *(12,13)*, the techniques are commercially available and straightforward to use, and they enable amplification of signals *(12)*.

The utility of (strept)avidin–biotin immunohistochemistry relies solidly on either of two assumptions: that endogenous biotin is below the detection limit of the assay *(7,14)* or that endogenous biotin is present homogenously throughout tissues and contributes only a uniform background label *(15)*. Unfortunately, in some tissues, both assumptions are false. Biotin is detectable in many tissues as a cofactor for carboxylases required in oxidative metabolism *(16,17)* and is often found with a heterogeneous distribution. These problems have been well recognized in renal and hepatic tissues, where copious endogenous biotin may lead to the false-positive interpretation of immunohistochemical experiments and clinical histopathological reports *(14,18–21)*. Importantly, strategies have been developed to minimize the detection of endogenous biotin, and the need to complete proper control experiments when using (strept)avidin–biotin detection systems in these tissues is frequently emphasized *(14,22,23)*.

However, the potential for endogenous biotin to contribute to false-positive interpretations in neural tissues has not often been appreciated. In addition to the biotin content noted in homogenates of rat brain *(1,24)*, several lines of evidence suggested to us that biotin may be present in significant quantities in brain. These included the neuron-specific expression of the biotin recycling enzyme biotinidase *(25–27)*, the presence of active carrier-mediated biotin transport mechanisms in brain *(19,28–32)*, the expression of biotin in several neural types from non-mammalian organisms *(15,33–38)*, and the clinical finding that patients with disorders of biotin metabolism often present with neurological indicators *(39–43)*. These facts prompted us to complete a thorough examination of the distribution of biotin in rat brain *(44)*.

Here, we report the detailed methodology used to identify the distribution of endogenous biotin in rat brain with three different techniques: a monoclonal antibody raised against biotin, a streptavidin-based detection system, and an avidin-based detection system. We also describe the testing and successful verification of strategies for combating endogenous biotin in neural tissue that have been previously reported in other tissues, including (strept)avidin–biotin blocking kits, prolonged fixation, and omission of Triton X-100 from the procedure. Finally we describe how to assess the non-specific (non-biotin) binding of (strept)avidin to brain tissue and report several strategies that were tested to reduce this problem. The reader is referred to our original paper for photomicrographs of all results *(44)*. Cumulatively our results indicate a significant potential for false-positive immunolabel when avidin- or streptavidin-based detection systems are used for the study of specific neural cell types unless specific strategies are implemented.

2. Materials

All chemicals are obtained from Sigma (St. Louis, MO, USA) unless otherwise noted.

2.1. Solutions

1. Phosphate buffer (PB) solution. Prepare a stock solution of 0.5 M dibasic phosphate (di-sodium hydrogen orthophosphate; Na_2HPO_4; EM Science, Gibbstown, NJ, USA) by slowly adding 70.98 g of Na_2HPO_4 to 1 l double distilled water (ddH_2O). Prepare a stock solution of 0.5 M monobasic phosphate (sodium dihydrogen phosphate; NaH_2PO_4; EM Science) by slowly adding 69.00 g NaH_2PO_4 to 1 l ddH_2O. Both stock solutions may require gentle warming and stirring (Na_2HPO_4 clumps if added too quickly). Prepare PB by combining 300 ml dibasic phosphate stock solution, 100 ml monobasic phosphate stock solution, and 1600 ml ddH_2O. Adjust pH to 7.4 with NaOH. If 0.5 M phosphate-buffered saline (PBS) solution is desired, then add 58.44 g NaCl per 2 l PB. PBS is best used for the avidin–biotin complex (ABC) reaction.
2. Paraformaldehyde (PARA) (4%) in PB. PARA is a carcinogen—complete all procedures in fume hood with all necessary precautions. Add 20 g PARA to 450 ml 0.1 M PB. Stir and heat solution to 65°C. Add 10 N NaOH dropwise until the solution clears. Allow solution to cool to room temperature and pH to 7.4 with 10 N HCl. Make volume up to 500 ml by adding 0.1 M PB. Filter with suction and store at 4°C (maximum 1 week).
3. Working solution. 3% normal serum (Jackson ImmunoResearch, West Grove, PA, USA), 0.1 % Triton X-100, balance PB (*see* **Note 1**).
4. Poly-D-lysine. Prepare 4 mg/ml poly-D-lysine in ddH_2O (MW >300,000). Apply a large drop (~15 µL) to the end of one microscope slide, then evenly smear across the slide using the edge of a second slide (*see* **Note 2**).
5. Anti-fade medium. Dissolve 50 mg p-phenylenediamine in 5 ml PB; sonication may be required. Add 45 mg glycerol; adjust pH to 10 with NaOH (*see* **Note 3**).

Store at −20°C in the dark and in a properly sealed container to prevent oxidation. The solution should be a light amber colour—do not use if solution has darkened.

2.2. Intracardial Perfusion

1. Sodium pentobarbital (65 mg/kg; MTC Pharmaceuticals, Cambridge, ON, USA).
2. Dissection tray and pins.
3. Surgical scissors (World Precision Instruments, Sarasota, FL, USA).
4. Surgical clamps (World Precision Instruments).
5. Large volume syringe (Becton Dickinson & Co., Franklin Lakes, NJ, USA).
6. Blunted syringe tip (Becton Dickinson & Co.).
7. Thin flexible tubing to connect syringe to syringe tip (Tygon, Akron, OH, USA).
8. PB, 250 ml.
9. PARA, 100 ml.

2.3. Tissue Slicing

1. Vibratome.
2. PB.
3. Tissue wells (Becton Dickinson & Co.) (*see* **Note 4**).
4. 'Fine point round' brushes (*see* **Note 5**).

2.4. Fluorescent Labelling (see Note 6)

1. Monoclonal α-biotin antibody (1:500).
2. AlexaFluor-488 conjugated donkey anti-mouse IgG (1:1000; Molecular Probes, Eugene, OR, USA).
3. Streptavidin-Cy3 (1:1500).
4. Streptavidin–biotin blocking kit (Vector Laboratories, Burlingame, CA, USA).
5. PB.
6. Working solution.
7. Microscope slides: 25 × 75 × 1 mm 'precleaned colorfrost microslides' and 22 × 50 mm (1 ounce) 'micro cover glasses' cover slips (VWR Scientific, West Chester, PA, USA) (slides coated with poly-D-lysine).
8. Anti-fade medium.
9. Nail polish (*see* **Note 7**).

2.5. Avidin–Biotin Complex

1. Vector ABC Elite HRP Kit (Vector Laboratories).
2. 3,3′-diaminobenzidine (DAB) reaction: In 20 ml PBS add 20 µl of 3 mg/ml glucose oxidase, 40 µl of 0.2 g/ml ammonium chloride, and 160 ul of 0.25 g/ml D+ glucose; mix with one DAB tablet and let dissolve for approximately 10 min. DAB is a carcinogen. Filter solution with a 0.2-µm filter (Pall Corporation, Ann Arbor, MI, USA).

3. PBS.
4. Working solution.
5. Avidin–biotin blocking kit (Vector Laboratories).
6. Microscope slides: 25 × 75 × 1 mm 'precleaned colorfrost microslides' and 22 × 50 mm (1 ounce) 'micro cover glasses' cover slips (both available from VWR Scientific) (slides coated with poly-D-lysine).
7. Ethanol.
8. Xylene.
9. Entellan (VWR Scientific).

2.6. Animals

1. Procedures must be completed following approval of, and in accordance with, institutional and national guidelines (*see* **Note 8**).

3. Methods

Here, we describe the procedures necessary to detect primary antibodies (and thus tissue antigens) with (i) secondary antibodies conjugated to fluorescent labels, (ii) biotinylated secondary antibodies and a tertiary fluorescent label conjugated to (strept)avidin, and (iii) biotinylated secondary antibodies with a tertiary complex of avidin, biotin, horseradish peroxidase, and then DAB to identify the peroxidase. Modifications of these procedures to identify the distribution of endogenous biotin within rat brain are described, and changes to the procedures to minimize the detection of endogenous biotin are discussed.

3.1. Intracardial Perfusion and Post-Fixation

1. Deeply anaesthetize rat with subcutaneous injection of 65 mg/kg sodium pentobarbital.
2. Place rat in supine position on dissection tray and affix limbs to tray with dissection pins.
3. Make incision from the level of the diaphragm to the top of the chest, cutting through the sternum to expose the heart. Use surgical clamps to hold open the exposed chest cavity.
4. Intracardial perfusion is accomplished with a large volume syringe connected to the blunted tip of an 18-gauge needle via thin flexible tubing. The blunted syringe tip should be gently inserted into the apex of the left ventricle and fixed in place with a surgical clamp.
5. Cut the inferior vena cava.
6. Slowly (1–2 ml/sec) perfuse at least 250 ml of room temperature PB through the heart (*see* **Note 9**).

7. Slowly perfuse at least 100 ml of room temperature 4% PARA through the heart (*see* **Note 10**).
8. Remove brain, place in 4% PARA at room temperature for 1 h, and leave overnight in 4% PARA at 4°C.

3.2. Tissue Slicing

1. Wash brain in PB for 2 h.
2. Cut free-floating 30- to 40-μm sections in ice-cold PB on vibratome.
3. Transfer slices into PB and store at 4°C until use (within 24–48 h) or transfer into 2% PARA and store at 4°C until use (1 week) (*see* **Note 11**).

3.3. Immunohistochemistry with Fluorophores (see Note 12)

1. Wash tissue sections 3 × 20 min in working solution; for tissue stored in PARA wash 4 × 60 min in PB prior to working solution washings (*see* **Note 13**).
2. Add primary antibody to working solution and apply to tissue for 24 h at 4°C (*see* **Note 14**).
3. Wash tissue sections 3 × 20 min in working solution to remove any unbound primary antibody.
4. Add secondary antibody to working solution (either a fluorophore-conjugated antibody or a biotinylated antibody) and apply to tissue for 24 h at 4°C. If secondary antibody is coupled to a fluorophore, then tissue must be stored in the dark (good to wrap tissue wells in tinfoil) and all subsequent procedures done with as little light as possible. If the secondary antibody is biotinylated, then the (strept)avidin–biotin blocking reaction should be completed just prior to this step (*see* **Subheading 3.6.** and **Fig. 1**).
5. Wash tissue sections 3 × 20 min is PB.
6. Mount tissue labelled with fluorophores on poly-D-lysine-coated slides.
7. Coverslip with anti-fade medium.
8. Seal cover slips to slides with nail polish and store at –20°C.
9. Tissue sections reacted with the biotinylated secondary antibody must be subsequently reacted with an avidin- or streptavidin-conjugated marker. Fluorophore-conjugated avidin or streptavidin are applied to the tissue at 4°C for 4 h in the dark and then subsequently mounted and stored as described above.

3.4. ABC–HRP Reactions

1. Wash tissue in 0.3% H_2O_2 in PBS for 10 min to quench endogenous peroxidase activity (*see* **Note 13**).
2. Wash 1 × 10 min in PBS [Given the high isoelectric point (pI = 10) of avidin, PB for reactions employing the Vector Elite kit should include 0.5 M NaCl (PBS) to minimize non-specific interactions *(6)*].
3. Wash 3 × 20 min in working solution.
4. Add primary antibody to working solution and apply to tissue for 24 hours at 4°C.

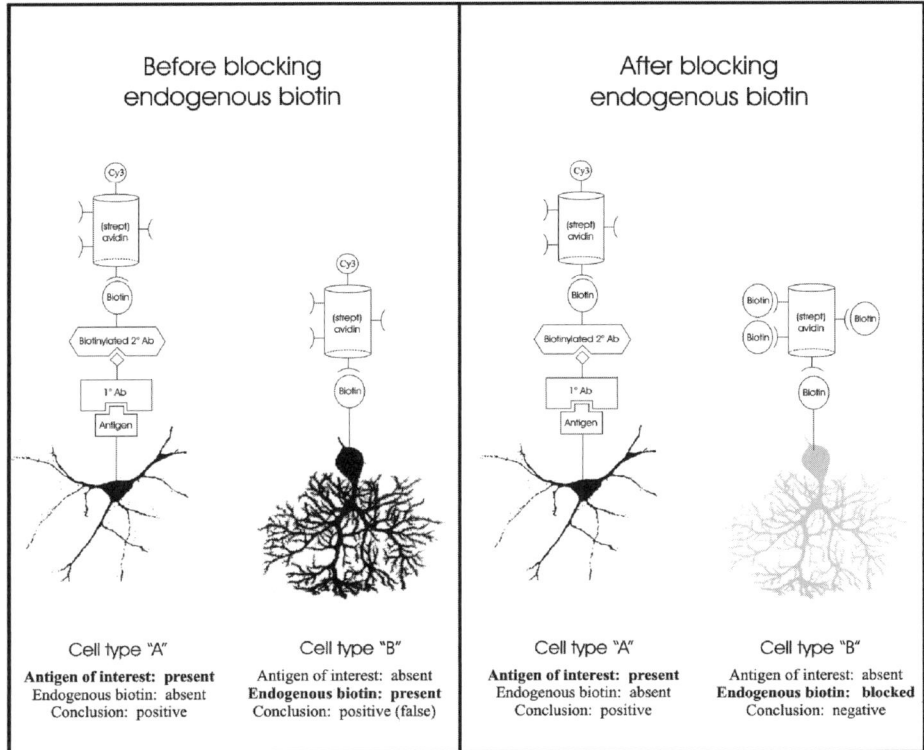

Fig. 1. Avidin- and streptavidin-conjugated markers detect both biotinylated antibodies and endogenous biotin in rat brain neurons potentially leading to false-positive results. For illustrative purposes, consider the cerebellar Golgi cell (Cell Type 'A') which does not contain detectable amounts of endogenous biotin, and the cerebellar Purkinje cell (Cell Type 'B') which contains readily detectable amounts of biotin in the cell body. (**A**) Although only the Golgi cell expresses the antigen of interest, use of the primary antibody → biotinylated secondary antibody → (strept)avidin-conjugated marker reaction sequence will label both the Golgi cell and the Purkinje cell. (**B**) After blocking endogenous biotin by applying excess unlabelled (strept)avidin and excess free biotin, application of the same 3-step reaction series illustrated in (**A**) now identifies only the Golgi cell. Note that the cerebellar Golgi and Purkinje cells used in this figure were filled with biotin through a patch pipette during electrophysiological recordings and subsequently labelled with a streptavidin-conjugated marker following histological processing, clearly showing the elaborate structural detail that can be realized using (strept)avidin–biotin technology in this slightly different context.

5. Wash tissue sections 3 × 20 min in PBS to remove any unbound primary antibody, and then wash 3 × 20 min in working solution.
6. Add biotinylated secondary antibody to working solution and apply to tissue for 24 h at 4°C.
7. Wash tissue sections 3 × 20 min in PBS.
8. Perform ABC reaction as indicated by Vector Laboratories.
9. Wash 3 × 10 min in PBS.
10. Apply DAB solution to tissue for approximately 10 min. The progress of the reaction should be monitored and terminated if the tissue begins to darken.
11. Wash 3 × 10 min in PBS.
12. Mount tissue sections on poly-D-lysine-coated slides and allow to dry overnight at room temperature.
13. Dehydrate slides in 70% alcohol for 5 min, 90% alcohol for 5 min, and 100% alcohol for 5 min.
14. Process slides twice through 100% xylene for 5 min each.
15. Apply one drop of Entellan to slide and cover slip.
16. Lay slides flat and allow them to dry.

3.5. Endogenous Biotin in the Rat CNS

We used the above methodologies to detect biotin in rat brain neurons *(44)*. The first approach was a conventional immunolabelling procedure where we examined the distribution of biotin using a monoclonal α-biotin antibody visualized with a secondary antibody coupled to AlexaFluor-488. Secondly, we used streptavidin-Cy3 to probe for biotin in a single step. Finally, we used the ABC of the Vector Elite kit with visualization of the horseradish peroxidase (HRP) conjugate by DAB. Endogenous biotin expression was identified in specific nuclei, with complete agreement in the distribution of neuronal label between the three cytochemical techniques, as revealed in co-labelling experiments.

Several cell types in the cerebellar motor system were prominently labelled for biotin. These included Purkinje cells (cell bodies with a limited distribution into the proximal dendrites), neurons of the deep cerebellar nuclei, neurons within the pontine nuclei, and neurons of the red nucleus. Neurons of the brainstem auditory system were also prominently labelled. For instance, labelling was noted in neurons of the spiral ganglia, the dorsal and ventral cochlear nuclei, the superior olivary nucleus, the MNTB, and neurons from the nucleus of the lateral lemniscus. More rostral auditory structures were not labelled. A few forebrain structures were also labelled, including neurons of the substantia nigra pars reticulata and neurons of the lateral mammillary nucleus. Within the hippocampus, interneurons were weakly positive for endogenous biotin with only a handful identified in any given tissue section, as previously reported *(18)*. Glial cells over the entire rostral-caudal extent of the spinal trigeminal tract contained endogenous biotin. Glial cells from other major tract systems, including the optic tract and the anterior commissure, were also biotin-positive. For the fluorescence experiments, MNTB neurons were the most intensely labelled, additional cell types of the auditory system as well as Purkinje cells were of intermediate intensity, and the cell types in the forebrain were labelled the weakest. For the avidin–biotin HRP complex experiments, all cell types were labelled equivalently. This result is attributed to a saturation of signal amplification, ultimately terminated by steric hindrance, which renders the Vector Elite ABC technique unusable for relative comparisons in staining intensity *(45)*. The distribution of endogenous biotin that we report closely matches the distribution of the biotin-recycling enzyme biotinidase in rat brain, although some discrepancies were noted *(25)*. It remains unclear why certain cell types may have greater, and thus detectable, quantities of biotin compared to other immuno-negative cell types. Significant concentrations of biotin are unlikely related to unique functional purposes, energetic demands, or cell size *(36)* (*see* **Notes 15** and **16**).

3.6. (Strept)avidin–Biotin Blocking Procedure

Blocking the binding of (strept)avidin to endogenous biotin was described shortly after the introduction of (strept)avidin–biotin technology to immuno-histochemistry *(14)*. The (strept)avidin–biotin blocking procedure is illustrated in **Fig. 1** and identifies the pitfalls associated with labelling endogenous biotin. Blockade is based on the following principles: (i) the high-affinity interaction between avidin and biotin, (ii) each (strept)avidin molecule is a tetramer with one biotin-binding site per subunit, and (iii) each biotin molecule has a single (strept)avidin-binding site. When tissue is incubated in excess free unlabelled (strept)avidin, all endogenous sites that may bind (strept)avidin (e.g., biotin, lectins, and hydrophobic/ionic interactions) are saturated. When the tissue is then incubated in free biotin, all available sites on immobilized (strept)avidin are blocked by biotin. This creates a closed system in which only experimentally applied (strept)avidin and biotinylated compounds can interact (*see* **Note 17**). The efficacy of a commercially available kit for blocking the detection of endogenous biotin by avidin- or streptavidin-conjugated markers is illustrated in **Fig. 2B, E and H**.

1. Prior to application of biotinylated primary or secondary antibodies, incubate tissue sections in unlabelled avidin solution or unlabelled streptavidin solution for 20 min (Vector Laboratories).
2. Wash 1 × 5 min in PB.
3. Incubate tissue in free biotin for 20 min.
4. Wash again 1 × 5 min in PB.
5. Proceed with immunohistochemical experiment as described above (*see* **Subheading 3.3.** or **3.4.**).

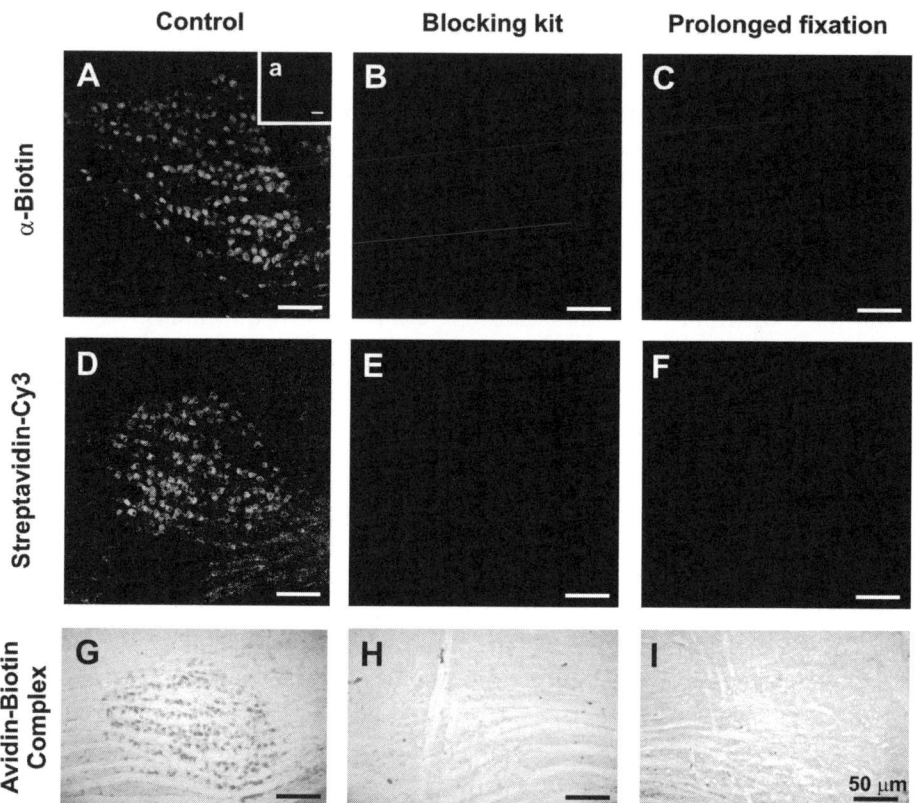

Fig. 2. Blocking endogenous biotin in the medial nucleus of the trapezoid body (MNTB). α-Biotin visualized with a secondary antibody conjugated to AlexaFluor-488 (**A**), streptavidin-Cy3 (**D**), and Vector Elite ABC (**G**) identifies identical distributions of endogenous biotin in neurons of the MNTB (**a**: secondary antibody with fluorophore only). Pre-absorption of α-biotin with excess free biotin (**B**), incubation with a commercially available streptavidin and biotin blocking kit (**E**), or incubation with a commercially available avidin and biotin blocking kit (**H**) all eliminated binding to endogenous biotin molecules. Additional post-fixation of the tissue destroyed or occluded endogenous biotin sites resulting in negative staining for all three techniques (**C**, **F**, **I**). Note that sections of the MNTB are devoid of endogenous biotin label when Triton X-100 is omitted from the working solution (not shown). Previous reports have indicated that (strept)avidin–biotin blocking kits are effective in blocking endogenous biotin in frozen tissue sections *(47)* and in paraformaldehyde-fixed sections but not in gluteraldehyde-fixed sections *(18)*.

3.7. Additional Strategies to Block Endogenous Biotin

The membrane permeabilizing detergent Triton X-100 is commonly used to facilitate the penetration of antibodies into tissues and across cell membranes. In fact, biotin, which is attached to intracellular carboxylases, is more readily identified in both rat stomach and frog brain when Triton X-100 is incorporated into immunohistochemical procedures *(34,46)*. We found that when Triton X-100 was removed from our working solutions, the α-biotin antibody, streptavidin-Cy3, and the ABC kit no longer identified endogenous biotin *(44)*. Previous reports have also suggested that PARA fixation blocks endogenous biotin detection *(18,47)*. We tested this and found that whereas 24-h PARA fixation did not block the detection of endogenous biotin, prolonged fixation (1 week) was effective (*see* **Fig. 2**) *(44)*. The use of long fixation times or the omission of Triton must be balanced against a potential decrease in the availability of the real antigen of interest.

3.8. Minimizing Non-Specific Tissue Interactions of Avidin and Streptavidin

Avidin and streptavidin share several properties: they have similar molecular weights, both are tetramers with one biotin-binding site per subunit, and both display an equal affinity for biotin *(10)*. However, there are several differences that merit consideration for immunohistochemical studies. Streptavidin, isolated from cultures of the bacterium *Streptomyces avidini*, has a weakly acidic pI, moves as an anion during electrophoresis, and is non-glycosylated *(10)*. Conversely, avidin, isolated from egg white, has a strongly basic pI *(9)*, moves strongly as a cation during electrophoresis, and is rich in glycoproteins *(10)*. The cationic and glycoprotein-rich properties of avidin are likely to contribute histochemically to non-specific ionic interactions with anionic materials in neural tissues *(48)*, such as the negatively charged glycosaminoglycans that are present on the surface of cells and within the intracellular matrix *(49)*. These concerns are not expected for the weakly anionic and non-glycosylated streptavidin.

The avidin-based ABC–HRP detection system used in our study consistently identified the same populations of biotin-positive neurons. However, there was a variable level of non cell-specific background labelling. This background label showed a pronounced rostral-caudal gradient, characterized by a low amount of label in the forebrain and a significantly greater density in the caudal brain (particularly the fibre tracts of the metencephalon and myelencephalon) *(44)*. This gradient was only weakly evident in sections treated with the anti-biotin antibody or streptavidin-Cy3 and appeared only as a restricted spotted or hazy distribution. A key potential source for non-specific interactions are negatively charged glycosaminoglycans present within the extracellular matrix that are found in close association with the surface of cells *(49–51)*. The exact composition of proteoglycans in the extracellular matrix also varies between brain regions *(50)*, potentially accounting for higher levels of background label detected by avidin conjugates in specific brain nuclei.

Non-specific binding of avidin to tissue is reportedly prevented by buffers of high ionic strength (e.g., 0.3–0.5 M NaCl) or alkaline pH (e.g., pH = 9.4) *(6,52,53)*. Hydrophobic and other protein–protein interactions can be blocked by adding 1–3% crystalline grade bovine serum albumin (BSA) or fetal calf serum to the working solution *(14)*. It is important to note that many commercially available antibody sera include BSA, which may reduce background labelling. Avidin can further bind to condensed chromatin *(54)* or to mast cells due to their heparin content *(53,55)*. Streptavidin, which binds to lectins, is reportedly suppressed with α-methyl-D-mannoside *(56)*. Commercially available blocking kits are expected to saturate both specific (biotin) and non-specific (strept)avidin-binding sites, and thus block cell-specific as well as background interactions.

We pursued the following strategies to block non-specific interactions:

1. Vary the ionic strength of the PB (0 M, 0.25 M and 0.5 M NaCl).
2. Raise the pH of the buffer to 9.4.
3. Include 1% BSA in the working solution.
4. Pre-incubate sections in 0.05 M α-methyl-D-mannoside.
5. Apply strept(avidin)–biotin blocking kits.

These changes may be implemented singly or in combinations. We found that pre-incubating the tissue in α-methyl-D-mannoside, including BSA in the working solution, or changing the ionic strength or pH of the buffer reduced but did not reliably eliminate all background label, with little effect on cell-specific biotin label. However, we noted that when HRP-labelled sections with low background were obtained, the PB was invariably of high ionic strength (0.5 M NaCl). The streptavidin–biotin blocking kit was found to block all cell-specific and non-specific interactions with tissue, whereas the avidin–biotin blocking kit blocked cell-specific labelling and only slightly reduced the non-specific interactions of the ABC kit with the tissue. Non-fat dry milk powder to block non-specific interactions is not advised *(57)*, as it has been shown to add additional protein bands in blotting experiments *(52)*. The use of dilute egg white solutions is ill-advised for this reason *(58)*, and because egg white contains avidin. Binding of biotinylated reagents to tissue is not expected, as endogenous avidin is restricted to birds and a few other oviparous vertebrates *(59)* (*see* **Note 18**).

3.9. Imaging

1. Tissue labelled with fluorescent markers should be imaged as soon as possible, as the intensity of the fluorescent signal fades with time whereas tissue labelled with the ABC kit (Vector Laboratories) will last for a substantial time.
2. Image processing is readily accomplished with Adobe Photoshop (Adobe Systems Inc., San Jose, CA, USA). Figures are readily prepared from these images in either Corel Draw (Corel, Ottawa, ON, USA) or Adobe Illustrator (Adobe Systems Inc.) (*see* **Note 19**).

4. Conclusion

We have demonstrated that avidin- and streptavidin-conjugated markers identify the same distribution of label in the rodent CNS as a monoclonal antibody directed against biotin. It is thus necessary to consider the potential for false-positive reactions with endogenous biotin in these regions when using (strept)avidin–biotin technology. The simplest way to sidestep this problem is to label antigens only with secondary antibody conjugates. If the amplification of a signal is required through the use of avidin or streptavidin conjugates, the need to complete proper controls is paramount. The most important of these is applying the avidin or streptavidin conjugate alone to probe for the presence of endogenous biotin or other sources of non-specific binding, especially in the rat hindbrain. Strategies for reducing non-specific association by avidin or streptavidin, such as those employed in the present report, are strongly encouraged to limit false-positive results attributable to endogenous biotin.

5. Notes

1. Blocking serums are chosen according to the host producing the secondary antibody (e.g., use donkey serum if secondary antibody was raised in donkey, goat serum if secondary antibody was raised in goat, and so on).
2. If the tissue tends to float off the slide during mounting, then the concentration of poly-D-lysine should be increased. However, with higher concentrations of poly-D-lysine, the tissue will be permanently stuck to the slide, and thus mounting errors (e.g., small folds in the tissue) will be uncorrectable.
3. Adjust pH over several minutes as pH changes very slowly in this highly viscous solution.
4. For small tissue sections use standard 24-well tissue culture Falcon Multiwell dishes. For larger tissue sections, standard 12-well culture dishes are better. To ensure proper exposure of tissue to the reagents, the tissue within each well should not overlap with one another.
5. 'Fine point round' brushes (available from most artwork stores) are ideal to transfer free-floating tissue slices between wells during the multiple steps of the

immunohistochemical procedures. Brushes should be carefully marked as to the primary antibody with which they have been in contact and only used in future experiments with the same primary antibody to avoid possible contamination.

6. For co-labelling experiments requiring two fluorophores, the absorption and emission spectra must be carefully considered to ensure that the dyes do not 'bleed-through' into one another (e.g., that the two dyes are not activated by similar wavelengths of light). Proper dye selection is complemented by proper configuration of the microscope filter sets. We use antibodies/reagents conjugated to the green label AlexaFluor-488 or the red labels AlexaFluor-555 or Cy3 (very similar to AlexaFluor-555) (all available from Molecular Probes), as they have non-overlapping absorption spectra.

7. Matte finish, light coloured, viscous nail polish works best. Avoid coloured nail polish that can autofluoresce and thus create a background signal during illumination.

8. In reptilian and amphibian CNS, the amount of biotin is known to fluctuate significantly between animals and over time and has been postulated to be related to the source of animals, environmental or seasonal factors, or even variability in diet *(15,34)*.

9. Quality of perfusion can initially be gauged by noting that the paws and tail of the rat, which are typically a pinkish colour, should become quite pale. Sufficient PB has been perfused when the fluid emerging out of the chest cavity has transitioned from blood red to a clear (PB) colour. Two additional strategies may be pursued to facilitate the removal of blood from the vascular system of larger rats: add 1 i.u. heparin (Leo Pharma Inc., Thornhill, ON, USA) per ml PB prior to perfusion or inject approximately 50 i.u. heparin into the vena cava and allow to circulate for approximately 30 s prior to initiating perfusion.

10. High-quality PARA perfusion has likely been accomplished if the body and neck of the rat are stiff.

11. PARA will diffuse from tissue over several days, and thus fixed tissue slices stored in PB for more than a few days will become unusable as the antigens degrade.

12. Note that each antigen and each antibody has unique procedural requirements that must be determined empirically. Thus, optimal labelling is achieved by considering the type of fixation (e.g., PARA vs. glutaraldehyde), the slicing procedure (free-floating vs. cryostat), numbers and durations of washings, duration of time that the antibodies are left to react with the tissue, and so on.

13. Use gentle agitation throughout all procedures to ensure thorough washing or exposure of tissue sections to antibodies or markers.

14. Several sections should be processed in parallel to the antibody-reacted sections and divided into several groups: one group receives an antibody known to label a specific population of cells [positive control—a background cell marker such as microtubule-associated protein 2 (BD Biosciences Canada, Mississauga, ON, USA) is useful here], one antibody that does not label the tissue of interest (negative control), one group that receives no antibody (primary

antibody omission control—this step should identify endogenous biotin if a (strept)avidin-conjugated marker is employed), and one group that receives the primary antibody after it has been pre-absorbed with an excess of its antigen.

15. A limited number of cell types in the CNS of invertebrates and non-mammalian vertebrates are positive for biotin. For invertebrates, biotin has been identified in the lobster giant axon *(36)* and in neurosecretory cells of the insect *Manduca sexta (37)*. In non-mammalian vertebrates, biotin has been identified in a subset of neurons from the turtle spinal cord *(15)*, in axons and synaptic boutons from numerous brain regions in frog *(34)*, in salamander and goldfish retina *(33,38)*, and in two nuclei following song learning in birds *(35)*. In mammals, the myelin fraction of rat brain *(45)*, and a few glial and neuronal cell types, including cerebellar Purkinje cells of pigs that are fed a biotin-rich diet *(60)*, are positive for endogenous biotin. Two earlier reports addressed the presence of endogenous biotin in the rat CNS, but only within a narrow region of forebrain spanning from the anterior hippocampus to anterior corpus callosum. These studies identified a few oligodendrocytes in tract systems *(61)* and intermittent labelling of neurons in hippocampus *(18)*.

16. 2D gel electrophoresis studies indicate at least 25 biotin-containing proteins in mammalian tissues *(62–64)*. Four of the principal bands have been described and correspond to pyruvate carboxylase, propionyl carboxylase, acetyl CoA carboxylase, and the α-subunit of methylcrotonyl-CoA carboxylase. The identities of the remaining 20 plus bands have not been determined but likely match several members of an extensive list of biotin-containing proteins *(17,65)*. Avidin–biotin, streptavidin–biotin, and α-biotin antibody approaches cannot discriminate between any of the biotin-containing proteins, free biotin, or biocytin *(48)*.

17. A similar approach termed the 'avidin chase' is used in radio-immunoimaging for human patients to reduce diagnostic interference and protect healthy biotin-rich organs from radiation poisoning when avidin- or streptavidin-conjugated markers are used to visualize biotinylated anti-tumor antibodies *(66–68)*.

18. In addition to false-positive results attributable to the above factors, false-negative results are possible with (strept)avidin–biotin detection systems. The ABC can become so large that steric hindrance may block the recognition of tissue antigens and create a false-negative result *(45)*. Furthermore, the large size of the ABC can inhibit penetration into tissue *(69)*. This fact may explain the success of Triton X-100 in unmasking endogenous biotin sites. Endogenous biotin has also been recognized as a significant immunohistochemical problem following other antigen retrieval techniques (e.g., heat-induced epitope retrieval), but can be blocked by avidin–biotin incubations *(70,71)*.

19. Image adjustments with the image processing software should be limited to intensity levels or brightness/contrast. Identical camera settings and level-contrast adjustments should be used for pairs of control and test photomicrographs during image processing; the sections themselves should be taken from

the same brain. A histologist blinded to the treatment conditions should always be consulted to assess the distribution of label, or the classification of a label as positive.

Acknowledgments

This work was supported by the Canadian Institutes for Health Research and an Alberta Heritage Foundation for Medical Research Scientist Award to RWT. BEM was supported by a Canadian Institutes for Health Research Canada Graduate Scholarship, a Natural Sciences and Engineering Research Council of Canada Postgraduate Fellowship, a Killam Trust Scholarship, and a Steinhauer Doctoral Award. The authors gratefully acknowledge the histological expertise of M. Kruskic and L. McKay.

References

1. Gyorgy, P. (1939) *J Biol Chem* **131,** 733–44.
2. du Vigneaud, V., Melville, D. B., Gyorgy, P., and Rose, C. S. (1940) *Science* **92,** 62–3.
3. Gyorgy, P., and Rose, C. S. (1940) *Science* **92,** 609.
4. Gyorgy, P., Melville, D. B., Burk, D., and du Vigneaud, V. (1940) *Science* **91,** 243–45.
5. Heitzmann, H., and Richards, F. M. (1974) *Proc Natl Acad Sci USA* **71,** 3537–41.
6. Guesdon, J. L., Ternynck, T., and Avrameas, S. (1979) *J Histochem Cytochem* **27,** 1131–9.
7. Hsu, S. M., Raine, L., and Fanger, H. (1981) *J Histochem Cytochem* **29,** 577–80.
8. Eakin, R. E., Snell, E. E., and Williams, R. J. (1941) *J Biol Chem* **140,** 535–43.
9. Woolley, D. W., and Longsworth, L. G. (1942) *J Biol Chem* **142,** 285–90.
10. Chaiet, L., and Wolf, F. J. (1964) *Arch Biochem Biophys* **106,** 1–5.
11. Green, N. M. (1975) *Adv Protein Chem* **29,** 85–133.
12. Wilchek, M., and Bayer, E. A. (1990) *Avidin-Biotin Technology*, Academic Press, San Diego.
13. Diamandis, E. P., and Christopoulos, T. K. (1991) *Clin Chem* **37,** 625–36.
14. Wood, G. S., and Warnke, R. (1981) *J Histochem Cytochem* **29,** 1196–204.
15. Berkowitz, A. (2002) *Brain Res* **938,** 98–102.
16. Dakshinamurti, K., and Chauhan, J. (1989) *Vitam Horm* **45,** 337–84.
17. Moss, J., and Lane, M. D. (1971) *Adv Enzymol Relat Areas Mol Biol* **35,** 321–442.
18. Wang, H., and Pevsner, J. (1999) *Cell Tissue Res* **296,** 511–6.
19. Bhagavan, H. N., and Coursin, D. B. (1970) *J Neurochem* **17,** 289–90.
20. Iezzoni, J. C., Mills, S. E., Pelkey, T. J., and Stoler, M. H. (1999) *Am J Clin Pathol* **111,** 229–34.
21. Seidman, J. D., Abbondanzo, S. L., and Bratthauer, G. L. (1995) *Int J Gynecol Pathol* **14,** 331–8.

22. Bratthauer, G. L. (1994) *Methods Mol Biol* **34,** 175–84.
23. Bratthauer, G. L. (1999) *Methods Mol Biol* **115,** 203–14.
24. Koivusalo, M., Elorriaga, C., Kaziro, Y., and Ochoa, S. (1963) *J Biol Chem* **238,** 1038–42.
25. Heller, A. J., Stanley, C., Shaia, W. T., Sismanis, A., Spencer, R. F., and Wolf, B. (2002) *Hear Res* **173,** 62–68.
26. Oizumi, J., and Hayakawa, K. (1989) *Biochim Biophys Acta* **991,** 410–4.
27. Oizumi, J., and Hayakawa, K. (1990) *Arch Biochem Biophys* **278,** 381–5.
28. Spector, R., and Mock, D. M. (1988) *Neurochem Res* **13,** 213–9.
29. Spector, R., and Mock, D. (1987) *J Neurochem* **48,** 400–4.
30. Lo, W., Kadlecek, T., and Packman, S. (1991) *J Nutr Sci Vitaminol* **37,** 567–72.
31. Chiang, G., and Mistry, S. (1974) *Proc Soc Exp Biol Med* **146,** 21–4.
32. Sander, J. E., Packman, S., and Townsend, J. J. (1982) *Neurology* **32,** 878–80.
33. Bhattacharjee, J., Nunes Cardozo, B., Kamphuis, W., Kamermans, M., and Vrensen, G. F. (1997) *J Neurosci Methods* **77,** 75–82.
34. Eisner, S., Harris, E., Galoyan, S. M., Lettvin, J. Y., and Scalia, F. (1996) *J Comp Neurol* **368,** 455–66.
35. Johnson, F., Norstrom, E., and Soderstrom, K. (2000) *Brain Res Dev Brain Res* **120,** 113–23.
36. Ma, P. M. (1994) *J Comp Neurol* **341,** 567–79.
37. Ziegler, R., Engler, D. L., and Davis, N. T. (1995) *Insect Biochem Molec Biol* **25,** 569–74.
38. Kurenny, D. E., Thurlow, G. A., Turner, R. W., Moroz, L. L., Sharkey, K. A., and Barnes, S. (1995) *J Comp Neurol* **361,** 525–36.
39. Wolf, B. (2001) *The Metabolic and Molecular Bases of Inherited Disease* (Valle, D., Ed.), pp. 3935–62, McGraw-Hill, New York.
40. Wolf, B., and Feldman, G. L. (1982) *Am J Hum Genet* **34,** 699–716.
41. Wolf, B., Grier, R. E., Allen, R. J., Goodman, S. I., and Kien, C. L. (1983) *Clin Chim Acta* **131,** 273–81.
42. Wolf, B., Heard, G. S., Weissbecker, K. A., McVoy, J. R., Grier, R. E., and Leshner, R. T. (1985) *Ann Neurol* **18,** 614–7.
43. Suchy, S. F., McVoy, J. S., and Wolf, B. (1985) *Neurology* **35,** 1510–11.
44. McKay, B. E., Molineux, M. L., and Turner, R. W. (2004) *J Comp Neurol* **473,** 86–96.
45. Sternberger, L. A., and Sternberger, N. H. (1986) *J Histochem Cytochem* **34,** 599–605.
46. Satoh, S., Tatsumi, H., Suzuki, K., and Taniguchi, N. (1992) *J Histochem Cytochem* **40,** 1157–63.
47. Hsu, S. M. (1990) *Avidin-Biotin Technology* (Wilchek, M., and Bayer, E. A., Eds.), Vol. 184, pp. 357–63, Academic Press, Inc., New York.
48. Savage, M. D., Mattson, G., Desai, S., Nielander, G. W., Morgensen, S., and Conklin, E. J. (1992) *Avidin-Biotin Chemistry: A Handbook*, Pierce Chemical Company, Rockford, IL.

49. Hook, M., Kjellen, L., Johansson, S., and Robinson, J. (1984) *Annu Rev Biochem* **53,** 847–69.

50. Yamaguchi, Y. (2000) *Cell Mol Life Sci* **57,** 276–89.

51. Celio, M. R., Spreafico, R., De Biasi, S., and Vitellaro-Zuccarello, L. (1998) *Trends Neurosci* **21,** 510–5.

52. Clark, R. K., Tani, Y., and Damjanov, I. (1986) *J Histochem Cytochem* **34,** 1509–12.

53. Jones, C. J., Mosley, S. M., Jeffrey, I. J., and Stoddart, R. W. (1987) *Histochem J* **19,** 264–8.

54. Heggeness, M. H. (1977) *Stain Technol* **52,** 165–9.

55. Bussolati, G., and Gugliotta, P. (1983) *J Histochem Cytochem* **31,** 1419–21.

56. Naritoku, W. Y., and Taylor, C. R. (1982) *J Histochem Cytochem* **30,** 253–60.

57. Duhamel, R. C., and Johnson, D. A. (1985) *J Histochem Cytochem* **33,** 711–4.

58. Miller, R. T., and Kubier, P. H. T. (1997) *Appl Immunohistochem Mol Morphol* **5,** 63–6.

59. Hertz, R., and Sebrell, W. H. (1942) *Science* **96,** 257.

60. Cooper, K. M., Kennedy, S., McConnell, S., Kennedy, D. G., and Frigg, M. (1997) *Res Vet Sci* **63,** 219–25.

61. LeVine, S. M., and Macklin, W. B. (1988) *Brain Res* **444,** 199–203.

62. Banks, R. E., Craven, R. A., Harnden, P. A., and Selby, P. J. (2003) *Proteomics* **3,** 558–61.

63. Chandler, C. S., and Ballard, F. J. (1985) *Biochem J* **232,** 385–93.

64. Chandler, C. S., and Ballard, F. J. (1986) *Biochem J* **237,** 123–30.

65. Wood, H. G., and Barden, R. E. (1977) *Annu Rev Biochem* **46,** 385–413.

66. Yao, Z., Zhang, M., Kobayashi, H., Sakahara, H., Nakada, H., Yamashina, I., and Konishi, J. (1995) *J Nucl Med* **36,** 837–41.

67. Kobayashi, H., Sakahara, H., Endo, K., Yao, Z. S., Toyama, S., and Konishi, J. (1995) *Eur J Cancer* **31A,** 1689–96.

68. Rusckowski, M., Fogarasi, M., Fritz, B., and Hnatowich, D. J. (1997) *Nucl Med Biol* **24,** 263–8.

69. Beltz, B. S., and Burd, G. D. (1989) *Immunocytochemical Techniques: Principles and Practice*, Blackwell Scientific Publications, Cambridge.

70. Bussolati, G., Gugliotta, P., Volante, M., Pace, M., and Papotti, M. (1997) *Histopathology* **31,** 400–7.

71. Rodriguez-Soto, J., Warnke, R. A., and Rouse, R. V. (1997) *Appl Immunohistochem Mol Morphol* **5,** 59–62.

11

Pseudo-Immunolabelling With the Avidin–Biotin–Peroxidase Complex due to the Presence of Endogenous Biotin in the Retina

Willem Kamphuis and Jan Klooster

Summary

Immunodetection techniques are dependent on enzyme–protein conjugates for the visualization of antigen–antibody complexes. One of the most widely used is the avidin–biotin–peroxidase complex (ABC) method. However, treatment of certain tissues with ABC reagents alone may result in high background, which is indicative for the presence of endogenous biotin or biotinylated proteins. In goldfish and salamander retinal sections, we observed a distinct staining pattern, presumably through binding of avidin to endogenous biotin in Müller cells. These findings summon for caution in the application of detection systems based on biotinylated antibodies or biotinylated DNA probes.

Key Words: Immunocytochemistry; endogenous biotin; ABC method; goldfish; salamander; zebrafish; rat; biotin; PAP method; Müller cells.

1. Introduction

Biotin is a vitamin present in minute amounts in living cells acting as a coenzyme for different carboxylases (acetyl-CoA, propionyl-CoA, β-methylcrotonyl-CoA, and pyruvate), enzymes involved in fatty acid synthesis, amino acid catabolism, and gluconeogenesis (1). Western blot studies have revealed biotinylated proteins with molecular weights corresponding the biotinylated subunits of carboxylases (2,3). The endogenous biotin distribution in rat kidney cells corresponds with that of a specific mitochondrial probe, and electron microscopic immunohistochemistry confirmed labelling of the mitochondrial matrix (4). Biotin is widely distributed in mammalian tissues

From: *Methods in Molecular Biology, vol. 418: Avidin-Biotin Interactions, Methods and Applications*
Edited by: R. J. McMahon © Humana Press, Totowa, NJ

and is present in relative high concentrations in liver, kidney, adipose tissue, and brain *(5–8)*.

The extremely high affinity of avidin, a 68-kDa glycoprotein, for biotin has been effectively exploited for a variety of purposes including quantitative enzyme immunoassays *(9)*, immunocytochemical localizations *(10,11)*, and in situ hybridizations *(12,13)*. Therefore, in any study using the avidin–biotin detection technology, the possibility of interference with endogenous biotin must be considered *(6–15)*.

In our ongoing immunocytochemical studies using the avidin–biotin–peroxidase complex (ABC) method, a clear example of such interference by endogenous biotin was encountered in the retina of goldfish and salamander *(16)* and to a lesser degree in zebrafish and rat. These results suggest that high levels of biotin are expressed in the retina of some species, putatively in the mitochondria of retinal Müller cells *(16)*. Because all tissues contain biotin, albeit at different amounts, performing proper controls is essential to exclude erroneous interpretation of staining results.

2. Materials

2.1. Animals and Tissue Preparation

1. Adult goldfish (*Carassius auratus*, 6–16 cm); salamander (*Amblystoma mexicanum*, 15–22 cm); zebrafish (*Danio rerio*, 2 cm); male Wistar rat (Harlan) 200–330 g.
2. MS222 (Sigma) and sodium-pentobarbital 60 mg/ml (Nembutal®).
3. Fixatives: 4% paraformaldehyde (PFA), or 4% PFA w/v/0.05% glutaraldehyde v/v, or 4% PFA w/v/15% saturated picric acid v/v, freshly made in 0.1 M sodium phosphate buffer (PB) at pH 7.2. All chemicals are obtained from Sigma or Merck unless stated otherwise.
4. Solutions of 15% sucrose and in 30% sucrose in PB at 4°C.
5. Tissue-Tek® (O.C.T.™ Compound; Sakura Finetek Europe) and dry ice.
6. Slides coated by dipping in 0.1 mg/ml poly-L-lysine (Fluka), dried for 2 h at room temperature.

2.2. ABC Staining

1. Phosphate-buffered saline (PBS), 0.1 M.
2. ABC (Vectastain *Elite* ABC kit, Vector Laboratories).
3. Diaminobenzidine (DAB; Sigma) 0.05%. One tablet of 10 mg is dissolved in 15 ml PBS, taken up in a syringe, expelled through a 0.2-μm filter, H_2O_2 (30%) is added to a final concentration of 0.01–0.02%, and used immediately.
4. Ethanol dehydation series (50% - 70% - 90% - 100% - 100% - 100% xylene - 100% xylene).
5. Embedding medium: Entellan® (Merck).

2.3. Immunocytochemistry

1. Normal rabbit serum and normal goat serum (Nordic Immunological Laboratories or Jackson Immuno Research).
2. Mouse monoclonal against biotin (Sigma; B7653). Diluted 1:1000 in PBS containing 2% bovine serum albumin (PBS/BSA).
3. Affinity purified peroxidase-labelled anti-biotin raised in goat (Vector Laboratories; SP-3010). Used at 1:2000 or 1:3000 dilution in PBS/BSA.
4. Mouse monoclonal against avidin (Sigma; A5680). Diluted 1:2000 or 1:4000 in PBS/BSA.
5. Secondary antibody: rabbit anti-mouse IgG (Nordic), diluted 1:100 in PBS/BSA, containing 1% NRS.
6. Mouse peroxidase-anti-peroxidase (mPAP) obtained from Nordic diluted 1:200 in PBS/BSA.
7. Rabbit-anti-mouse F(ab′)2-Cy3 labelled (Jackson Immuno Research) 1:700.
8. Vectashield embedding medium with DAPI (Vector Laboratories).

2.4. Western Blots

1. Homogenization buffer: 0.32 M sucrose, 1.0 mM EDTA, 0.25 mM dithiothreitol, 0.2 mM phenylmethylsulfonylfluoride (PMSF), pH 8.0.
2. Solubilization buffer: 20 mM Tris–HCl, 1% Triton X-100, 0.2 mM PMSF, pH 8.0.
3. Loading buffer: 62 mM Tris, 1 mM EDTA, 1% sodium dodecyl sulfate (SDS), 1% dithiothreitol.
4. Biotinylated SDS–polyacrylamide gel electrophoresis (PAGE) standards (BioRad).
5. SDS–PAGE, 10–15%, following standard procedures for protein electrophoresis (Mini Protean II system) and transfer to nitrocellulose membrane (Hybond ECL-Amersham) by electroblotting.
6. Blocking solution: 5% Blotting Grade Blocker Non-fat Drye Milk (BioRad) in Tris-buffered saline + 1% Tween-20.
7. Avidin–horseradish peroxidase (HRP) conjugate (BioRad) diluted to 1:750,000.
8. Chemiluminescence detection using an ECL-kit (Amersham Bioscience).

3. Methods

3.1. Tissue Preparation

1. Goldfish, salamander, and zebrafish are anaesthetized with MS222 and decapitated.
2. Rats (250–300 g) are euthanized with an overdose an overdose of sodium-pentobarbital.
3. Eyes are enucleated and hemi-sected by an encircling cut around the limbus, the cornea and lens removed, and the vitreous partially drained with absorbent tissue. Removal of the vitreous improves the reproducibility of the immunostaining quality.

4. Eyecups are immediately fixed by immersion fixation for 4–6 h at 4°C under constant agitation.

5. Eyecups are cryoprotected in 15% sucrose–PBS for 3 h, and in 30% sucrose–PBS overnight at 4°C. The duration of these steps may be shortened to the time needed for the tissue to sink to the bottom of the vial.

6. The tissue is embedded in Tissue-Tek® and quickly frozen on dry ice. For small pieces of tissue like the retina, a cap, cut off from a 1.5-ml Eppendorf tube, filled with Tissue-Tek® may serve as a freezing and storage container. After positioning the tissue in the cap filled with Tissue-Tek® sufficient to cover the tissue, the cap is placed in powdered dry ice. The cap is placed back on the tube, the tube is placed in a conventional rack, and stored at –80°C. This is a space-efficient way of tissue storage and also greatly facilitates the retrieval of specific samples from large series.

7. Cryosections are prepared cryostat (Leica CM3050). Transversely cut sections, 8–14 μm thick, are collected on coated slides, dried for 2 h at room temperature, and stored at –20°C in Parafilm-wrapped small plastic slide containers.

3.2. ABC–DAB Staining Procedure

1. Sections are thawed in closed containers to room temperature.

2. Sections are washed (4 × 10 min) in 0.1 M PBS to remove traces of fixative and sucrose. For the detection of biotinylated proteins in the tissue, sections are directly treated in ABC solution for 1 h. The two reagents of the kit (avidin DH and biotinylated peroxidase) are in a dilution of 1:50 in PBS and must stand for about 30 min before use. In this time, a matrix of many biotinylated enzyme molecules crosslinked by avidin is formed.

3. Sections are treated with 0.05% DAB + 0.01–0.03% H_2O_2 in PBS for 10–15 min at room temperature and constant agitation. The exact time needs to be optimized by performing a time series of incubations with increasing duration. When all conditions (tissue fixation, temperature, agitation, section thickness, and so on) are kept constant, a consistent staining intensity can be achieved. Visualization of the ABC-mediated staining by the Metal Enhanced DAB Substrate Kit (Pierce) results in sharper staining with less background.

4. Sections are washed in distilled water, dehydrated in ethanol series, cleared in xylene, and coverslipped in Entellan. A typical example is presented in **Fig. 1A**.

3.3. Immunocytochemistry for Detection of Endogenous Biotin

1. Sections are thawed in closed containers to room temperature.

2. Sections are washed (4 × 10 min) in PBS and additionally for 30 min in BSA/PBS.

3. Preincubation for 30 min in 5% normal rabbit serum to block non-specific binding of the rabbit anti-mouse IgG used a subsequent step.

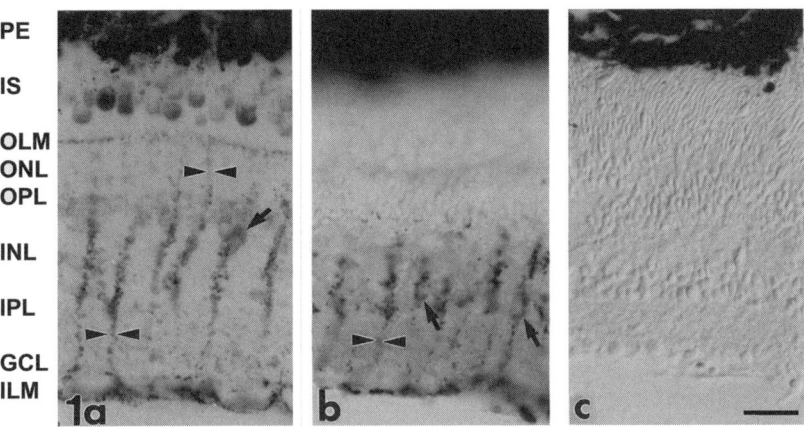

Fig. 1. Goldfish retina. Comparison of staining after avidin–biotin–peroxidase complex (ABC)–diaminobenzidine (DAB) procedure (**A**), with that of anti-biotin immunoreactivity, visualized by the peroxidase-anti-peroxidase (PAP) method (**B**), and anti-avidin–PAP immunoreactivity (**C**). Note the similarities in staining reactions between ABC–DAB (A) and anti-biotin (B), and the absence of staining reaction panel C (anti-avidin). GCL, ganglion cell layer; ILM, inner limiting membrane; INL, inner nuclear layer; IPL, inner plexiform layer; IS, inner segment ellipsoid; OLM, outer limiting membrane; ONL, outer nuclear layer; OPL, outer plexiform layer; PE, pigment epithelium. Arrows indicate diffuse perinuclear staining in panels A and B. Arrowheads indicate position of radial fibres of Müller cells. Bar represents 50 μm.

4. The sections are incubated in the diluted primary antibodies for 24–48 h (*see* **Note 3**). Slides are placed in closed Teflon trays together with soaked tissues to create moist conditions, at 4°C.
5. Excess of antibodies is removed by extensive washing in PBS (4 × 10 min) and in PBS/BSA (1 × 30 min).
6. Sections are incubated for 30 min with rabbit anti-mouse IgG, diluted 1:100 in PBS/BSA, containing 1% NRS.
7. The wash sequences in PBS, PBS/BSA are repeated and incubation with diluted mPAP for 1 h is carried out.
8. After washing in PBS (4 × 10 min), the sections are treated with DAB/H_2O_2, washed, dehydrated, and mounted. An example of the resulting staining is presented in **Fig. 1B**.
9. For immunofluorescent localization of biotin, rabbit anti-mouse (*see* **step 6**) is replaced by rabbit anti-mouse Cy3 incubated for 1 h. Ater washing, sections are coverslipped in Vectashield containing DAPI. For results in goldfish, zebrafish, and rat, *see* **Fig. 2** (*see* **Note 4**).

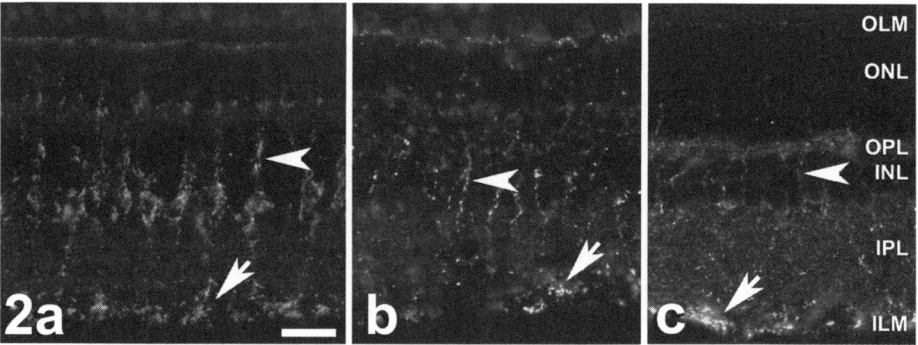

Fig. 2. Distribution of biotin immunoreactivity by immunofluorescence in the retina of goldfish (**A**), zebrafish (**B**), and rat (**C**). Arrowheads indicate position of radial fibres of Müller cells. Arrows indicate the typical end-feet morphology of Müller cells near the ILM. ILM, inner limiting membrane; INL, inner nuclear layer; IPL, inner plexiform layer; OLM, outer limiting membrane; ONL, outer nuclear layer; OPL, outer plexiform layer.

10. Alternatively, direct detection of endogenous biotin may be considered. Thawed sections are blocked with 5% NGS and incubated with goat anti-biotin conjugated with peroxidase. After washing, sections are treated with DAB + H_2O_2 in PBS for 10–15 min at room temperature.

3.4. Western Blotting for Endogenous Biotinylated Proteins

1. The eyecup is prepared as described in **Subheading 3.1.** The retina is isolated from the pigment epithelium and homogenized in the homogenization buffer on ice.
2. The homogenate is centrifuged (15,000 g, 30 min at 4°C).
3. The supernatant is retrieved and the pellet is solubilized in solubilization buffer for 1 h at 4°C. Unsolubilized proteins are removed by centrifugation and the two supernatant fractions are pooled.
4. Different amounts of protein (0.5–20 µg) are mixed with loading buffer and boiled for 5 min before loading on the gel. Biotinylated marker proteins are used as molecular mass standards. Conventional methods for Western blotting of protein samples are followed to fractionate total protein by 10–15% SDS–PAGE and subsequent transfer to nitrocellulose membrane by electroblotting.
5. After blocking the blot in 5% blocking protein in TBS-T, the blot was incubated with avidin–HRP.
6. HRP activity was visualized with chemiluminescence. An example of the results is shown in **Fig. 3**.

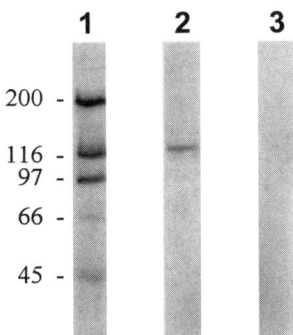

Fig. 3. Tissue homogenate of goldfish retina was subjected to electrophoresis and blotted to nitrocellulose membrane. After transfer to nitrocellulose membrane, the presence of biotinylated protein was visualized by incubation with an avidin–horseradish peroxidase (HRP) conjugate, followed by chemiluminescence detection of the HRP activity. Lane 1, biotinylated molecular weight standards, sizes (in kDa) are shown on the left. Lane 2, protein homogenate (0.5µg) of goldfish retina. A distinct band corresponding to a size of approximately kDa is revealed. Lane 3, preincubation with anti-biotin followed by avidin–HRP incubation. The detection of the band visualized in lane 2 is completely blocked (*see* **Notes 1** and **2, Table 1**).

4. Notes

1. The presence of biotin or biotinylated proteins is commonly detected when performing a control for background staining with ABC reagents only. When staining is observed, it is essential to routinely include, in all further studies, a blocking step. Blocking of endogenous biotin is effectively accomplished by a pre-incubation with an avidin D-containing solution *(15)* (*see* **Table 1**).
2. Around 1997, several papers reported interference with endogenous biotin. Some suppliers have put special 'blocking kits' and protocols on the market (Vector Laboratories, SP-2001). However, it is essential to test the quality of the blocking step carefully and to add controls for this to each experiment, especially when new fixatives or new tissue samples are studied.
3. Confirmation that the ABC staining pattern is indeed related to the presence of endogenous biotin can be obtained through additional studies with antibodies directed against biotin or with peroxidase-labelled antibody against biotin (*see* **Fig. 1 B**). In principle, endogenous avidin could also underlie the ABC-mediated background. Immunostaining against avidin in the goldfish retina was negative (*see* **Fig. 1C**).
4. The use of modern fluorescent Cy dyes, with increased sensitivity, circumvents the problems of endogenous biotin. Using this technique, we could show an extensive pattern of labelled punctae with a Muller cell-like pattern in goldfish

Table 1
Detection (+), Partial Suppression (±), and Complete Supression/Absence (–) of Endogenous Staining Pattern in Goldfish and Salamander Retina

Treatment	Goldfish	Salamander
ABC–DAB	+	+
DAB	–	–
Avidin–ABC–DAB	–	–
Avidin~HRP–DAB	–	–
Anti-biotin–rabbit anti-mouse–mouse PAP–DAB	+	+
Anti-biotin~PO–DAB	+	+
Anti-avidin–rabbit anti-mouse–mouse PAP–DAB	–	–
Anti-avidin–ABC–DAB	+	±
Anti-biotin–ABC–DAB	±	±

ABC, avidin–biotin–peroxidase complex; DAB, diaminobenzidine solution containing 0.01–0.03% H_2O_2; ~HRP, horseradish peroxidase labelled; PAP, peroxidase-anti-peroxidase; ~PO, peroxidase labelled.

retina (*see* **Fig. 2A**) but also in the zebrafish (*see* **Fig. 2B**) and rat retina (*see* **Fig. 2C**). In the latter two species, ABC revealed only faint staining, indicating that fluorescence-mediated detection of biotin is in fact more sensitive compared to ABC detection.

5. Further insight in the ABC-mediated staining pattern can be obtained by a series of blocking experiments. (i) The putative binding sites of endogenous biotin are blocked by anti-biotin antibodies, followed by ABC and DAB treatment. Similar incubation with anti-avidin did not reduce the discrete ABC-mediated staining pattern. (ii) Blockage of endogenous biotin by avidin followed by ABC and DAB yielded no staining. (iii) Results of the various sets of experiments carried out to characterize the nature of ABC staining in goldfish and salamander retina are summarized in **Table 1.**

6. **Figure 3** shows the result of biotin immunostaining with avidin–HRP on blotted retinal proteins of goldfish, revealing a distinct band at approximately 120 kDa which may correspond with the molecular size of pyruvate carboxylase of 125 kDa (*2*). Preincubation with the monoclonal antibody against biotin resulted in complete suppression of this band.

7. Tyramide-biotin signal amplification [TSA; Perkin-Elmer (NEN)] techniques may also suffer from interference with endogenous biotin. This technology uses HRP antibodies to catalyze the deposition of biotin-labelled tyramide, resulting in the deposition of numerous biotin labels in the section to be visualized by chromogenic or fluorescence techniques. Endogenous biotin may lead to a false-positive signal.

Acknowledgments

The work described in this chapter is partly based on the work of Dr. J. Bhattacharjee carried out during his stay at the NORI in 1996. Despite our best efforts, we have not been able to track down the current address of Dr. Bhattacharjee.

References

1. Dakshinamurti K. and Chauhan J. (1990) Nonavidin biotin-binding proteins. *Methods Enzymol.* **184,** 93–102.
2. Praul C. A., Brubaker K. D., Leach R. M.,and Gay C. V. (1998) Detection of endogenous biotin-containing proteins in bone and cartilage cells with streptavidin systems. *Biochem. Biophys. Res. Commun.* **247,** 312–314.
3. Ruggiero F. P. and Sheffield J. B. (1998) The use of avidin as a probe for the distribution of mitochondrial carboxylases in developing chick retina. *J. Histochem. Cytochem.* **46,** 177–183.
4. Hollinshead M., Sanderson J., and Vaux D. J. (1997) Anti-biotin antibodies offer superior organelle-specific labeling of mitochondria over avidin or streptavidin. *J. Histochem. Cytochem.* **45,** 1053–1057.
5. Dakshinamurti K. and Mistry S. P. (1963) Tissue and intracellular distribution of biotin-C-[14]00H in rats and chicks. *J. Biol. Chem.* **238,** 294–296.
6. Wang H. and Pevsner J. (1999) Detection of endogenous biotin in various tissues: novel functions in the hippocampus and implications for its use in avidin-biotin technology. *Cell Tissue Res.* **296,** 511–516.
7. Moss J. and Lane M. D. (1971) The biotin-dependent enzymes. *Adv. Enzymol. Relat. Areas Mol. Biol.* **35,** 321–442.
8. Wood H. G. and Barden R. E. (1977) Biotin enzymes. *Annu. Rev. Biochem.* **46,** 385–413.
9. Hsu S. M., Raine L., and Fanger H. (1981) Use of avidin-biotin-peroxidase complex (ABC) in immunoperoxidase techniques: a comparison between ABC and unlabeled antibody (PAP) procedures. *J. Histochem. Cytochem.* **29,** 577–580.
10. Wilchek M. and Bayer E. A. (1988) The avidin-biotin complex in bioanalytical applications. *Anal. Biochem.* **171,** 1–32.
11. Hsu S. M., Raine L., and Fanger H. (1981) Use of avidin-biotin-peroxidase complex (ABC) in immunoperoxidase techniques: a comparison between ABC and unlabeled antibody (PAP) procedures. *J. Histochem. Cytochem.* **29,** 577–580.
12. Chevalier J., Yi J., Michel O., and Tang X. M. (1997) Biotin and digoxigenin as labels for light and electron microscopy in situ hybridization probes: where do we stand? *J. Histochem. Cytochem.* **45,** 481–491.
13. Parkin R. K., Boeckh M. J., Erard V., Huang M. L., and Myerson D. (2005) Specific delineation of BK polyomavirus in kidney tissue with a digoxigenin-labeled DNA probe. *Mol. Cell. Probes* **19,** 87–92.

14. Bussolati G., Gugliotta P., Volante M., Pace M., and Papotti M. (1997) Retrieved endogenous biotin: a novel marker and a potential pitfall in diagnostic immuno-histochemistry. *Histopathology* **31,** 400–407.

15. Wood G. S. and Warnke R. (1981) Suppression of endogenous avidin-binding activity in tissues and its relevance to biotin-avidin detection systems. *J. Histochem. Cytochem.* **29,** 1196–1204.

16. Bhattacharjee J., Nunes C. B., Kamphuis W., Kamermans M., and Vrensen G. F. (1997) Pseudo-immunolabelling with the avidin-biotin-peroxidase complex (ABC) due to the presence of endogenous biotin in retinal Muller cells of goldfish and salamander. *J. Neurosci. Methods* **77,** 75–82.

12

Use of Synthetic Peptides for Identifying Biotinylation Sites in Human Histones

Gabriela Camporeale, Yap Ching Chew, Alice Kueh, Gautam Sarath, and Janos Zempleni

Summary

Posttranslational modifications of histones play an important role in the regulation of chromatin structure and, hence, gene regulation. Recently, we have identified a novel modification of histones: binding of the vitamin biotin to lysine residues in histones H2A, H3, and H4. Here, we describe a procedure to identify those amino acids that are targets for biotinylation in histones. Briefly, the following analytical sequence is used to identify biotinylation sites: (i) short peptides (<20 amino acids in length) are synthesized chemically; amino acid sequences in the peptides are based on the sequence in a given region of a given histone; (ii) peptides are incubated with biotinidase or holocarboxylase synthetase to conduct enzymatic biotinylation; and (iii) biotin in peptides are probed using streptavidin peroxidase. Amino acid substitutions (e.g., lysine-to-alanine substitutions) in synthetic peptides can be used to corroborate identification of biotinylation sites.

Key Words: Biotinylation; histones; chromatin; biotinidase; holocarboxylase.

1. Introduction

Histones are small proteins (11–22 kDa) that mediate the folding of DNA into chromatin. Five major classes of histones have been identified in eukaryotic cells: histones H1, H2A, H2B, H3, and H4 (1). DNA is wrapped around octamers of core histones, each consisting of one H3-H3-H4-H4 tetramer and two H2A-H2B dimers, to form the nucleosomal core particle. Each nucleosomal core particle binds about 146 base pairs of DNA (1). Nucleosomal assembly is completed by one molecule of histone H1 associating with the DNA that connects two nucleosomal core particles. Nucleosomes are stabilized

From: *Methods in Molecular Biology, vol. 418: Avidin-Biotin Interactions, Methods and Applications*
Edited by: R. J. McMahon © Humana Press, Totowa, NJ

by electrostatic interactions between negatively charged phosphate groups in DNA and positively charged ε-amino groups (lysine residues) and guanidino groups (arginine residues) in histones.

Histones consist of a globular C-terminal domain and a flexible N-terminal tail *(1)*. The N-terminal tail of histones protrudes from the nucleosomal surface; lysines, arginines, serines, and glutamates in the N-terminal tail are targets for acetylation, methylation, phosphorylation, ubiquitination, poly (ADP-ribosylation), and sumoylation *(1–5)*. These modifications play important roles in regulating processes such as transcriptional activation or silencing of genes and DNA repair.

Recently, a novel covalent modification of eukaryotic histones has been identified, that is, biotinylation of lysine residues *(6,7)*. All the major classes of histones are targets for biotinylation *(6)*. The following two biotinyl histone transferases have been identified in humans: biotinidase *(8)* and holocarboxylase synthetase *(9)*. These two enzymes utilize distinct substrates: biotinylation of histones by biotinidase depends on biocytin (biotinyl-ε-lysine) and biotinylation of histones by holocarboxylase synthetase depends on biotin and ATP.

Biotinylation of histones appears to be involved in gene silencing *(10)*, cell proliferation *(6,9)*, and DNA repair or apoptosis *(10,11)*. The following observations are consistent with the hypothesis that histone biotinylation is important for human health. First, biotinylation of the lysine residue in position 12 in histone H4 decreases rapidly in response to double-stranded DNA breaks *(11)*. This suggests that alterations in the biotinylation pattern of histones might be an early signaling event in response to DNA damage. Second, mutations of the genes encoding biotinidase *(12–14)* and holocarboxylase synthetase *(15)* are fairly common *(16,17)*. Fibroblasts from individuals with mutated holocarboxylase synthetase are severely deficient in histone biotinylation *(9)*. In analogy, mutated biotinidase is not capable of catalyzing biotinylation of histones *(8)*. Future study may reveal abnormal patterns of gene silencing *(10)*, cell proliferation *(6,9)*, and DNA repair or apoptosis *(10,11)* in afflicted individuals.

We have developed a procedure to identify those amino acids that are targets for biotinylation in histones *(7)*. Briefly, the following analytical sequence was used to identify biotinylation sites: (i) short peptides (<20 amino acids in length) were synthesized chemically; amino acid sequences in the peptides were based on the sequence in a given region of a given histone; (ii) peptides were incubated with biotinidase or holocarboxylase synthetase to conduct enzymatic biotinylation; and (iii) biotin in peptides was probed using streptavidin peroxidase. Amino acid substitutions (e.g., lysine-to-alanine substitutions) in synthetic peptides were used to corroborate identification of biotinylation sites *(7)*. In addition, amino acid modifications (e.g., acetylation of lysines) in peptides were used to investigate the cross-talk between biotinylation of histones and other known modifications of histones *(7)*. In this chapter, we provide protocols for the identification of biotinylation sites in histones by using synthetic peptides. These protocols have been used successfully to identify biotinylation sites in histones H4 *(7)*, H3 *(18)*, and H2A (unpublished findings).

Identification of biotinylation sites is a pre-requisite for generating site-specific antibodies to biotinylated histones. Such antibodies are invaluable tools (i) to study the cross-talk among modifications of histones, for example, biotinylation and acetylation of lysine residues *(7)*; (ii) to investigate biotinylated histones by using immunocytochemistry; and (iii) to investigate roles for biotinylation of histones in the regulation of transcriptional activity of genes by using chromatin immunoprecipitation assays.

2. Materials

2.1. Peptide Synthesis

1. Peptide synthesizer (Visiprep-DL vacuum Manifold, cat. no. 57044, Sigma-Aldrich, St. Louis, MO, USA).
2. Lyophilizer.
3. Glass Pasteur pipettes; pipette bulbs.
4. Filtration tubes, 3 mL, with polyethylene frits (cat. no. 57024, Sigma-Aldrich).
5. N-fluoren-9-ylmethoxycarbonyl (Fmoc)-amino acids, for example, N-α-Fmoc-L-alanine. Store desiccated at −80°C; bring to room temperature before opening the container.
6. 5-(4′-Aminomethyl-3′, 5′-dimethoxyphenoxy)-valeric acid (PAL) resin (ABI Inc., Foster City, CA, USA).
   The following reagents are hazardous; wear protective clothing and store reagents in a fume hood.
7. N, N-dimethylformamide (DMF, ≥99% purity).
8. O-(benzotriazol-1-yl)-N,N,N′,N′-tetramethyluronium tetrafluoroborate (TBTU) (16.1%) in DMF (w/v); stable for 5 days at room temperature.
9. N-ethyldiisopropylamine (DIPEA, ≥99% purity) (17.4%) in DMF (v/v); stable for 5 days at room temperature (1.74 mL DIPEA/10 mL DMF).
10. Piperidine (≥ 99% purity) (20%) in DMF (v/v); stable for 5 days at room temperature (*see* **Note 1**).
11. Methylene chloride, ≥99% purity.
12. Deblocking solution: 96% trifluoroacetic acid (≥99% purity), 2% thioanisole (≥99% purity), 1% triisopropylsilane (≥99% purity), 1% distilled water; can be stored for up to 12 h.
13. Tert-butyl-methyl-ether (≥99% purity).

2.2. Peptide Quantification

1. Cysteine-HCl monohydrate in distilled water at concentrations of 0–2.8 mM (calibration curve).

2. 1.0 M Tris buffer, pH 8.2, at 25°C, containing 0.02 M ethylenediaminetetraacetic acid (EDTA); store at 4°C.
3. 5,5′-Dithio-bis-2-nitrobenzoic acid (DTNB), 0.01 M, in methanol; prepare immediately before use.
4. Plate reader equipped with a 405-nm filter.

2.3. Enzymatic Biotinylation of Peptides Using Plasma Biotinidase

1. Biotinyl-ε-L-lysine (biocytin), 750 μM, 99% purity; store at –20°C in small aliquots.
2. Human plasma; store at –20°C in small aliquots.
3. Tris buffer, 50 mM, pH 8.0 at 37°C; store at 4°C.
4. Synthetic peptides (1 μg/μL) (*see* **Subheading 2.1.**).

2.4. Enzymatic Biotinylation of Peptides Using Biotinidase from Mouse Liver and Spleen

2.4.1. Preparation of Tissue Extract

1. SZ 22 tissue grind tube plus pestle (homogenizer; Kotes Glass Company, Vineland, NJ, USA).
2. Sonicater (Aquasonic™ 250T, cat. no. 21811-890, VWR Scientifics, West Chester, PA, USA).
3. Mouse tissue.
4. Homogenization buffer: 0.25 M sucrose, 50 mM Tris–HCl, 5 mM reduced glutathione, 1 mM disodium EDTA, pH 7.9 at 37°C; store at 4°C *(19)*.
5. Buffer A: 10 mM N-(2-hydroxyethyl) piperazine-N′-(2-ethane sulfonic acid) (HEPES), pH 7.9, 10 mM KCl, 0.1 mM EDTA, 0.1 mM ethylene glycol-bis(2-aminoethylether)-N,N,N′,N′-tetraacetic acid (EGTA), 1 mM dithiothreitol (DTT), 0.5 mM phenylmethanesulfonyl fluoride (PMSF) *(20)*.
6. Buffer B: 20 mM HEPES, pH 7.9, 0.4 M NaCl, 1 mM EDTA, 1 mM EGTA, 1 mM DTT, 1 mM PMSF *(20)*.
7. Igepal-630, 10%, in distilled water (v/v).

2.4.2. Enzymatic Biotinylation of Peptides

1. 750 μM Biocytin, 99% purity.
2. 50 mM Tris buffer, pH 8.0 at 37°C; store at 4°C.
3. Mouse homogenate.
4. Synthetic peptides (1 μg/μL) (*see* **Subheading 2.1.**).

2.5. Streptavidin Blots

1. Western blot unit, for example, the Invitrogen XCell Mini-Cell and Blot Module (Invitrogen Inc., Carlsbad, CA, USA).

2. Tricine SDS loading buffer, 2× (Invitrogen Inc.).
3. Tricine polyacrylamide gels, 16% (Invitrogen Inc.).
4. Tricine SDS running buffer, 1× (Invitrogen Inc.).
5. Chromatography filter paper (cat. no. 05-714-4, Fisher, Pittsburgh, PA, USA).
6. Transfer solution: 4% 25× Tris–Glycine transfer buffer (Invitrogen Inc.) and 20% methanol in distilled water.
7. Methanol, 100%.
8. Polyvinylidene difluoride (PVDF) membrane.
9. Dulbecco's phosphate-buffered saline (PBS).
10. Tween-20, 0.05%, in PBS (v/v) (Tween–PBS; use within 24 h).
11. Bovine serum albumin (BSA) (3%) in PBS (w/v); use within 12 h.
12. ImmunoPure® streptavidin horseradish peroxidase conjugate (Pierce, Rockford, IL, USA). Prepare stock solution (1 mg/mL) in distilled water, store at –20°C in 50 µL aliquots. Working solution: 0.25 µg/mL streptavidin in Tween–PBS (mix 10 µL streptavidin stock solution and 40 mL Tween–PBS).
13. SuperSignal® West Pico Chemiluminuscent Substrate (Pierce).
14. Rocking platform.
15. Autoradiography film.
16. Autoradiography cassette.
17. Autoradiography film developer.

3. Methods

3.1. Peptide Synthesis

1. Identify the amino acid sequence of a given histone by accessing GenBank (e.g., GenBank accession number NM_175054 for human histone H4). Select fragments based on histone sequence for synthesis. Typically, peptides containing up to 15 amino acids can be synthesized by using the method described here. Note that amino acids are conjugated in C-terminal to N-terminal direction in the nascent peptide. It is advisable to add a cysteine residue to the C terminus of each peptide to permit both peptide quantification (*see* **Subheading 3.2.**) and conjugation to immobilized carriers (*see* **Note 2**).
2. Before beginning a synthesis, weigh out all the amino acids needed. For each conjugation step, weigh out 25 mg of Fmoc-amino acids into 1.5-mL microfuge tubes. Store at –20°C until use.
3. For each peptide to be synthesized, add 8–10 µmoles of PAL resin sto a filtration tube with frit.
4. Close the valves on the vacuum manifold and add 2 mL of DMF to the filtration tubes containing PAL resin. Swell resin for at least 5 min.
5. Open the valves and wash the resin three times with 2 mL of DMF and drain completely.
6. Wash the resin with 0.25 mL of 20% piperidine in DMF, close the valve, and incubate the resin with another 0.25 mL aliquot of 20% piperidine in DMF for 5 min. Meanwhile, dissolve amino acids required for the specific coupling step

(depends on number of peptides being synthesized simultaneously) (from step 2) in 0.125 mL of 17.4% DIPEA in DMF and 0.125 mL of 16.1% TBTU in DMF. Dissolve amino acids for coupling reactions immediately before use.

7. Wash the resin three times with DMF: 3 mL, 1 mL, and 1 mL. Drain completely in-between washes.

8. Close the valve of the filtration tubes and add dissolved and activated amino acids prepared as described in **step 6**. Incubate at room temperature for 1 h (*see* **Note 3**).

9. Open the valve and wash the resin three times with DMF: 3 mL, 1 mL, and 1 mL. Drain completely in-between washes.

10. Repeat steps 5–8 as needed to add amino acids to the nascent peptide (*see* **Note 4**).

11. After the final DMF wash, rinse the filtration tube with 10 mL of methylene chloride drain tube and let dry for 24–48 h at room temperature (*see* **Note 5**).

12. To release peptides from the resin ("deblocking"), transfer the dry resin into a 2-mL microfuge tube. Save the filtration tubes with frits for use in step 13. Add 0.3–0.4 mL of deblocking solution to the microfuge tube by using a glass Pasteur pipette. Incubate the microfuge tubes at room temperature for 4 h.

13. Add 10 mL of cold *tert*-butyl methyl ether (chill bottle at –20°C for ~2 h) into a 15-mL polypropylene tube. Place the empty filtration tube from step 12 on top of the 15-mL polypropylene tube. Transfer the deblocking solution plus resin from the microfuge tube back into the filtration tube; collect the flow through (deblocking solution) in the cold ether. Wash the 2-mL microfuge tube three times with deblocking solution (0.35 mL/wash). Pass the deblocking solution through the filtration tube and collect all washes in the same polypropylene tube.

14. Chill the polypropylene tube at –20°C for 30 min. Centrifuge the tube (lightly capped) at maximum speed in a clinical centrifuge for 10 min (3000–5000 g). Discard the supernatant into a clearly marked glass waste bottle and dissolve the precipitated peptide in 2 mL of distilled water. Freeze at –80°C (*see* **Note 6**).

15. Lyophilize peptides twice: dissolve peptides in 2 mL of distilled water after first lyophilization, freeze at –80°C, and lyophilize for second time. Dissolve peptides in 1 mL of distilled water.

Note: Hazardous chemicals are used in steps 4–14. Perform these steps in a fume hood.

3.2. Peptide Quantification

This assay depends on the presence of sulfhydryl groups in synthetic peptides (*see* **Subheading 3.1., step 1**).

1. Prepare a dilution series of cysteine in distilled water (0–2.8 mM) for calibration.

2. In a 96-well plate, mix 20 μL aliquot of peptide solution with 178 μL of 1.0 M Tris buffer containing EDTA and 2 μL of DTNB in methanol. For calibration, substitute the peptide solution with cysteine standards. As a blank, use 200 μL 1.0 M Tris buffer containing EDTA.

3. Incubate plate for 10 min at room temperature.
4. Measure absorbance at 405 nm in a plate reader.
5. Remember to normalize values by taking into consideration the number of sulfhydryl groups in a given peptide.

3.3. Enzymatic Biotinylation of Peptides Using Biotinidase from Human Plasma

1. Adjust concentration of peptides (*see* **Subheading 3., steps 1** and **2**) to 1 µg/µL.
2. Mix 5 µL of peptide, 470 µL of 50 mM Tris buffer, 15 µL of human plasma (as a source of biotinidase), and 10 µL of 750 µM biocytin in a 1.5-mL microfuge tube.
3. Vortex gently and centrifuge briefly to collect the mixture at the bottom of the tube. Incubate at 37°C in a water bath for 45 min (*see* **Note 7**).

3.4. Enzymatic Biotinylation of Peptides by Biotinidase from Mouse Liver and Spleen Extracts

3.4.1. Preparation of Tissue Extract

1. Add 2 mL of homogenization buffer to about 0.5 g of mouse tissue and homogenize the sample on ice using a SZ 22 homogenizer.
2. Sonicate the sample on ice by using six 30-s bursts with a 1-min cooling period on ice after each burst.
3. Transfer the homogenized solution to a 15-mL polypropylene tube; add 10 mL of cold PBS.
4. Centrifuge the sample at 1500 *g* for 5 min at 4°C. Discard the supernatant.
5. Resuspend the pellet in 1 mL of cold PBS and transfer to a 1.5-mL microfuge tube.
6. Centrifuge the sample at maximal speed for 15 s in a microcentrifuge; discard the supernatant.
7. Resuspend the pellet by adding 400 µL ice-cold buffer A by gently pipetting up and down.
8. Incubate the sample on ice for 15 min.
9. Add 25 µL of 10% Igepal-630 and vortex vigorously for 10 s.
10. Centrifuge the homogenate for 30 s in a microcentrifuge at maximal speed.
11. Discard the supernatant and resuspend the nuclear pellet in 50 µL of ice-cold buffer B.
12. Shake the microfuge tube vigorously at 4°C for 15 min on a rocking platform.
13. Centrifuge the sample at 4°C for 5 min at 1200 *g*; freeze the supernatant in small aliquots at –80°C up to 1 month.

3.4.2. Enzymatic Biotinylation of Peptides

1. Mix 5 µL of peptide, 435 µL of 50 mM Tris buffer, 50 µL of nuclear extract, and 10 µL of 750 µM biocytin in a 1.5-mL microfuge tube.

2. Vortex gently and centrifuge briefly to collect the mixture at the bottom of the tube. Incubate at 37°C in a water bath for 45 min.

3.5. Streptavidin Blots (Protocol for Invitrogen System and Gels)

1. Mix 10 μL of sample (biotinylated peptide) with 10 μL of loading buffer (pre-warmed at 37°C). Heat at 95°C in a thermal cycler or heating block for 10 min.
2. Rinse the 16% SDS–tricine gel with distilled water, remove the white strip and comb from the gel, and place the gel into the western blot unit. Fill the wells in the gel with 1× tricine running buffer.
3. Load 15 μL of heated sample per well and fill the inner chamber of the western blot unit with 1× running buffer; be sure to cover the wells with buffer. Fill the outer chamber with 1× running buffer and resolve the peptides at 125 V for 90 min or until the dye front reaches the bottom of the gel.
4. While the gel is running, soak six pads in transfer buffer. Squeeze the submerged pads carefully to remove air bubbles. Cut two filter papers and PVDF membrane to match the size of the gel. Pre-wet the PVDF membrane in 100% methanol for about 10 s and soak it in transfer buffer. Soak the filter paper in transfer buffer just prior to use.
5. Break open the gel cassette and transfer peptides onto PVDF membrane by electroblotting at 25 V for 90 min.
6. Block the PVDF membrane in 3% BSA in PBS (w/v) for 1 h with gentle shaking (*see* **Note 8**).
7. Wash the PVDF membrane twice with PBS containing 0.05% Tween-20 for 5 min each.
8. Incubate PVDF membrane in streptavidin working solution for 1 h at room temperature on a rocking platform.
9. Wash membrane twice for 5 min with Tween–PBS, followed by one wash with PBS for 5 min.
10. Incubate membrane with chemiluminescence substrate for 5 min.
11. Quantify chemiluminescence by using autoradiography film (up to 5 min of exposure).

4. Notes

1. TBTU, DIPEA, and piperidine solutions are light sensitive. Prepare only as much as needed for the synthesis; keep the working solutions protected from light.
2. You may want to consider synthesizing a chemically biotinylated peptide as a positive control. This can be achieved by using a biotinylated amino acid derivative (e.g., biotin-ε-NH$_2$-L-lysine) (*see* **Subheading 3.1.**). Chemically biotinylated peptides generate a very intense signal in streptavidin blots and should be diluted 100–1000 fold.
3. To avoid contamination during synthesis, insert kimwipes on top of each filtration tube during the 1-h incubation.

4. Use the following procedure to interrupt peptide synthesis (e.g., overnight). After incubation of a given amino acid with PAL resin for 1 h (*see* **Subheading 3.1., step 8**), wash the filtration tube with DMF (*see* **Subheading 3.1., step 9**). Close the valve, add 2 mL of DMF, and seal the tube with parafilm to prevent evaporation. Store the sealed tube at room temperature for up to 24 h. To resume peptide synthesis, remove parafilm, open the valve, and drain the tube; continue as described in **step 6** (*see* **Subheading 3.1.**). To avoid unspecific reactions during synthesis, complete each peptide synthesis in no more than 5 days.

5. After the final wash with methylene chloride (*see* **Subheading 3.1., step 11**), the peptides are stable and can be stored for an extended period of time in a desiccator until being deblocked.

6. Deblocked peptides can be stored safely at –80°C until being lyophilized.

7. For storage of samples after enzymatic biotinylation, stop the reaction by adding an equal volume of 2× tricine SDS loading buffer. Heat 20 μL aliquots at 95°C for 10 min; freeze at –80°C.

8. After transferring peptides by electroblotting, PVDF membrane can be blocked in 3% BSA for up to 24 h (e.g., overnight) at 4°C.

Acknowledgments

This work was supported by NIH grants DK 60447 and DK 063945, by NSF EPSCoR grant EPS-0346476, by a grant from the Nebraska Tobacco Settlement Biomedical Research Enhancement Funds, and in part by NIH Grant Number 1 P20 RR16469 from the BRIN Program of the National Center for Research Resources.

References

1. Wolffe, A. (1998) *Chromatin*, Academic Press, San Diego, CA.
2. Fischle, W., Wang, Y., and Allis, C. D. (2003) *Curr. Opin. Cell Biol.* **15,** 172–83.
3. Jenuwein, T., and Allis, C. D. (2001) *Science* **293,** 1074–80.
4. Boulikas, T., Bastin, B., Boulikas, P., and Dupuis, G. (1990) *Exp. Cell Res.* **187,** 77–84.
5. Shiio, Y., and Eisenman, R. N. (2003) *Proc. Natl. Acad. Sci. U. S. A.* **100,** 13225–30.
6. Stanley, J. S., Griffin, J. B., and Zempleni, J. (2001) *Eur. J. Biochem.* **268,** 5424–29.
7. Camporeale, G., Shubert, E. E., Sarath, G., Cerny, R., and Zempleni, J. (2004) *Eur. J. Biochem.* **271,** 2257–63.
8. Hymes, J., Fleischhauer, K., and Wolf, B. (1995) *Biochem. Mol. Med.* **56,** 76–83.
9. Narang, M. A., Dumas, R., Ayer, L. M., and Gravel, R. A. (2004) *Hum. Mol. Genet.* **13,** 15–23.
10. Peters, D. M., Griffin, J. B., Stanley, J. S., Beck, M. M., and Zempleni, J. (2002) *Am. J. Physiol. Cell Physiol.* **283,** C878–84.

11. Kothapalli, N., and Zempleni, J. (2004) *FASEB J.* **18,** A103–4.

12. Swango, K. L., Demirkol, M., Huner, G., Pronicka, E., Sykut-Cegielska, J., Schulze, A., Mayatepek, E., and Wolf, B. (1998) *Hum. Genet.* **102,** 571–75.

13. Wolf, B., Jensen, K., Huner, G., Demirkol, M., Baykal, T., Divry, P., Rolland, M. O., Perez-Cerda, C., Ugarte, M., Straussberg, R., Basel-Vanagaite, L., Baumgartner, E. R., Suormala, T., Scholl, S., Das, A. M., Schweitzer, S., Pronicka, E., and Sykut-Cegielska, J. (2002) *Mol. Genet. Metab.* **77,** 108–11.

14. Moslinger, D., Muhl, A., Suormala, T., Baumgartner, R., and Stockler-Ipsiroglu, S. (2003) *Eur. J. Pediatr.* **162,** S46–9.

15. Yang, X., Aoki, Y., Li, X., Sakamoto, O., Hiratsuka, M., Kure, S., Taheri, S., Christensen, E., Inui, K., Kubota, M., Ohira, M., Ohki, M., Kudoh, J., Kawasaki, K., Shibuya, K., Shintani, A., Asakawa, S., Minoshima, S., Shimizu, N., Narisawa, K., Matsubara, Y., and Suzuki, Y. (2001) *Hum. Genet.* **109,** 526–34.

16. Wolf, B., and Heard, G. S. (1991) Biotinidase deficiency, in *Advances in Pediatrics* (Barness, L., and Oski, F., eds.), Medical Book Publishers, Chicago, IL, pp. 1–21.

17. Wolf, B. (1991) *J. Inherit. Metab. Dis.* **14,** 923–27.

18. Sarath, G., Kobza, K., Rueckert, B., Camporeale, G., Zempleni, J., and Haas, E. (2004) *FASEB J.* **18,** A103 [abstract].

19. Zempleni, J., Trusty, T. A., and Mock, D. M. (1997) *J. Nutr.* **127,** 1776–81.

20. Schreiber, E., Matthias, P., Muller, M. M., and Schaffner, W. (1989) *Nucleic Acids Res* **17,** 6419.

13

Detection and Quantitation of the Activity of DNA Methyltransferases Using a Biotin/Avidin Microplate Assay

Kirsten Liebert and Albert Jeltsch

Summary

The biotin–avidin microplate assay is a sensitive method to measure methylation of biotinylated oligonucleotide substrates by DNA methyltransferases (MTases). The methylation reaction is carried out in solution using [methyl-^{3}H]-AdoMet. Afterwards, the oligonucleotides are immobilized on an avidin-coated microplate, where the incorporation of [^{3}H]-labeled methyl groups into the DNA is stopped by addition of unlabeled AdoMet to the binding buffer. Separation of radioactively labeled DNA from unreacted AdoMet and enzyme is performed by washing steps. Subsequently, the radioactivity incorporated into the DNA is released by a nucleolytic digestion of the DNA. By liquid scintillation counting, the amount of DNA methylation can be determined. Advantages of the microplate assay are its high sensitivity which allows the detection of low amounts of DNA methylation, the efficient separation of reaction components resulting in a low background of radioactivity and a high accuracy ($\pm10\%$) and reliability. Furthermore, the assay is very convenient, fast and well suited for automation.

Key Words: DNA modification; DNA methyltransferase; biotin–avidin interaction; enzyme assay.

1. Introduction

DNA methylation is an important and essential modification of DNA observed in most prokaryotes and eukaryotes that has many biological functions. In prokaryotes, DNA methylation is used to coordinate DNA replication and cell cycle, to direct post-replicative mismatch repair and to distinguish between self and nonself DNA. In eukaryotes, it is involved in processes

From: *Methods in Molecular Biology, vol. 418: Avidin-Biotin Interactions, Methods and Applications*
Edited by: R. J. McMahon © Humana Press, Totowa, NJ

like gene regulation, maintenance of genome integrity, regulation of development and protection of genome against selfish DNA *(for review see* **refs.** *1–5)*. Erroneous DNA methylation causes different diseases including cancer *(6,7)*. DNA methyltransferases (MTases) catalyze the transfer of a methyl group to the N^6 position of adenine, N^4 position of cytosine or C^5 position of cytosine by using S-adenosyl-L-methionine (AdoMet) as a donor for an activated methyl group *(for review see* **refs.** *1* and *8)*. Because of its diverse biological functions, methods for analysis of this process are very important and deserve continuous refinement. Several assay systems have been developed to study the activity of DNA MTases *(for review see* **refs.** *9* and *10)* including digestion of the DNA by methylation-sensitive restriction enzymes *(11–13)*, bisulfite modification for detection of 5-methylcytosine *(14,15)* or the separation and quantitative determination of modified nucleosides by high pressure liquid chromatography (HPLC) *(16–18)*. Another class of methylation assays relies on the usage of AdoMet that carries a radioactive label on its methyl group that is transferred to the DNA by the MTase. These assays require the separation of methylated DNA and unused cofactor which can be achieved by (i) spotting the reaction mix onto a DE-cellulose filter sheet *(19)*, (ii) coupling of DNA to cellulose *(20)*, or (iii) thin layer chromatography *(21)*. Some years ago, we have introduced a biotin–avidin microplate format for this purpose *(22)*. This assay is a sensitive in vitro method to assay the activity of DNA MTases that allows measuring low amounts of DNA methylation in a fast, accurate way.

In the microplate assay described in this chapter *(see* **Fig. 1**), biotinylated oligonucleotide substrates are methylated using [methyl-^{3}H]-AdoMet. During the reaction, the radioactive label is transferred from the coenzyme to the DNA. After the methylation reaction, the methylated DNA must be separated from unreacted AdoMet and coenzyme bound to the enzyme. For this separation, the DNA is immobilized on the surface of an avidin-coated microplate. During this step, incorporation of [^{3}H] into the DNA is quenched by the addition of an excess of unlabeled AdoMet in the binding buffer. In the next step,

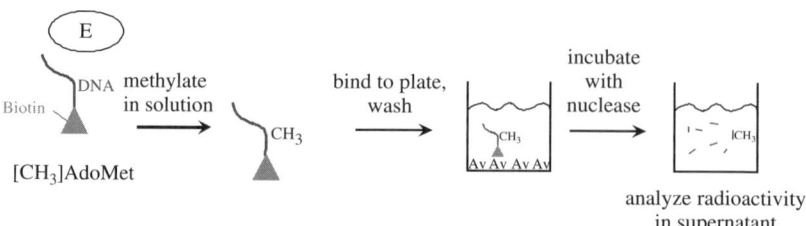

Fig. 1. Schematic drawing of the biotin/avidin microplate DNA methylation assay described in this chapter.

the free AdoMet and enzyme-bound AdoMet are removed by washing. In the end, the bound DNA is digested with a nonspecific nuclease to release the incorporated radioactivity. After the digestion, the radioactivity in the solution is measured by liquid scintillation counting to quantify the amount of methyl groups transferred to the DNA by DNA MTases.

This protocol has advantages in comparison to published protocols. First, the background of radioactivity is low, because of the efficient removal of unreacted AdoMet. Second, the detection of [^{3}H] by liquid scintillation counting is possible with high efficiency. Third, results can be accurately reproduced with small deviations of ±10%. Taken these points together, this assay is very well suited to detect methylation of DNA at a level of smaller than 0.1% of total methylation of the DNA. The assay provides quantitative data of high accuracy and reproducibility. Furthermore, it enables the usage of the microplate format to process many samples in parallel in a fast and inexpensive way. Therefore, the method is prone for automation and high-throughput approaches.

2. Materials

1. Microplates (e.g., E.I.A./R.I.A. plate, flat bottom, high binding, cat. no 9018, Costar Corp., Cambridge, MA, USA; or comparable product from any other manufacturer); store coated plates at 4°C.
2. Avidin (Sigma-Aldrich Laborchemikalien GmbH, Seelze, Germany); store at 4°C.
3. Unlabeled AdoMet (Sigma); 10 mM solution in 10 mM H_2SO_4; aliquot; store at –20°C.
4. 100 mM $NaHCO_3$ pH 9.6.
5. Phosphate Buffered Saline Tween-20 (PBST): 140 mM NaCl, 2.7 mM KCl, 4.3 mM Na_2HPO_4, 1.4 mM K_2HPO_4, 0.05% Tween 50, pH 7.2
6. Multichannel pipette.
7. Biotinylated oligonucleotide substrates; anneal and prepare a stock solution that is stored at –20°C.
8. Reaction buffer for the methylation reaction, composition depends on the enzyme studied.
9. [methyl-^{3}H]-AdoMet (3.22 TBq/mmol, NEN), store at –20°C in small aliquots.
10. Nonspecific endonuclease, for example, *Serratia marcescens* nuclease in 50 mM Tris/HCl pH 8, 5 mM $MgCl_2$; the enzyme is commercially available as Benzonase from Merck (Merck Biosciences GmbH, Bad Soden, Germany), DNaseI might be used as well.
11. HCl, 0.05 M.
12. Liquid Scintillator Solution Rotizint® Eco Plus (Carl Roth GmbH + Co. KG, Karlsruhe, Germany); Irritant.
13. Liquid scintillation counter and counter vials

3. Methods

1. Preparation: Coat the microplates with avidin; dissolve 1 µg avidin in 100 µl of 100 mM $NaHCO_3$, pH 9.6 (*see* **Note 1**), pipette 100 µl per well and incubate overnight at 4°C. Coated plates can be stored for 2 weeks at 4°C.
2. Before use, wash the wells five times with 200 µl PBST to remove unbound avidin (*see* **Note 2**).
3. Prepare the binding buffer consisting of 5 µl 10 mM unlabeled AdoMet in 35 µl PBST supplemented with 500 mM NaCl (*see* **Note 3**) and distribute it in each well of the microplate to quench the incorporation of [^{3}H] into the DNA after the methylation reaction.
4. Methylation reactions are carried out in 10–50 µl in a reaction tube. Typically, 0.1–10 µM biotinylated oligonucleotide and 1 nM–10 µM enzyme is used in a buffer adapted to the enzyme to be studied, for example, 100 mM HEPES ((4-(2-hydroxyethyl)-1-piperazineethanesulfonic acid)), pH 8.0, 50 mM NaCl, 1 mM EDTA (Ethylenediamine tetraacetic acid), 0.5 mM DTT (Dithiothreitol), 200 ng/µl bovine serum albumin in the presence of 0.75 µM labeled [methyl-^{3}H]-AdoMet (3.22 TBq/mmol, NEN) (*see* **Note 4**). Methylation reactions can be carried out at different temperatures. The reaction can be started by addition of enzyme, DNA or coenzyme. To measure a time course of methylation (*see* **Note 5**), remove aliquots of 1–5 µl from the reaction mixture at each time point and pipette them into the wells of the microplate that contain the binding buffer with an excess of unlabeled AdoMet (*see* **Note 3**) to quench the incorporation of [^{3}H] into the DNA. After the last time point, incubate the plate for 30 min at room temperature to allow binding of the oligonucleotide substrates to the avidin on the microplate.
5. Wash the wells five times with 200 µl PBST supplemented with 500 mM NaCl to remove the unreacted AdoMet and the enzyme (*see* **Notes 2** and **6**).
6. Digest the immobilized DNA by adding 0.7 µg *S. marcescens* nuclease (*see* **Note 7**) in 100 µl 50 mM Tris/HCl pH 8.0, 0.5 mM $MgCl_2$ per well and shake the microplate for 30 min at room temperature. As an alternative, release of the DNA could be achieved by adding 100 µl 0.05 M HCl per well (which disrupts the avidin–biotin interactions) and shaking the microplate for 30 min at room temperature (*see* **Note 8**).
7. After digestion, the whole solution is transferred from the wells into a scintillation vessel. Each sample is mixed with 2 ml Liquid Scintillator Solution and subjected to liquid scintillation counting to quantify the amount of methyl groups transferred to the DNA (*see* **Note 9**).

4. Notes

1. Wrong pH leads to a lower binding efficiency of avidin to the surface of the microplate.
2. Avoid scratching the bottom of the plate with the pipette tips. Tilting of the plate facilitates complete removal of avidin or washing buffer.
3. Prepare the binding buffer that contains an excess of unlabeled AdoMet to quench the incorporation of [^{3}H] into the DNA directly before using. Unlabeled AdoMet

should be stored in small aliquots in 10 mM H_2SO_4 and thawed only once to avoid degradation. Use high-salt buffer to prevent binding of the MTase to the DNA after stopping the reaction.

4. Labeled AdoMet should be stored at –20°C, aliquoted and thawed only once.

5. Carry out each measurement at least in duplicates.

6. Complete removal of the MTase and unreacted labeled AdoMet is very important for a low background of radioactivity. After each washing step, no buffer should be left in the wells.

7. The purification of *S. marcescens* nuclease was performed similarly as described by Friedhoff et al. *(23)*: The His_6-Nuclease fusion protein was expressed in TGE900 *Escherichia coli* cells. Protein overexpression was induced at a cell density of 0.5–0.6 at OD 600 nm by changing the temperature from 28 to 42°C, and the cells were grown for an additional 2 h. All following steps were carried out at 4°C. The cells were harvested by centrifugation (15 min and 3000 g) and washed with STE buffer [10 mM Tris/HCL (pH 8.0), 0.1 mM EDTA, 0.1 mM NaCl]. The cell pellet was resuspended in buffer A (10 mM Tris/HCl, pH 8.2), and the cells were disrupted by ultrasound. Cell debris were removed by centrifugation (1 h at 15,000 g), and the pellet was resuspended in buffer B (6 M urea, 10 mM Tris/HCl, pH 8.2, 10 mM imidazole) and kept overnight at 4°C on a shaker. After a new centrifugation (1 h at 15,000 g), the supernatant was applied onto a Ni-NTA column (Qiagen GmbH, Hilden, Germany) equilibrated with buffer B. The column was washed with 150 ml buffer B, eluted with buffer B containing 200 mM imidazole and collected in fractions. Fractions containing the nuclease were pooled and dialyzed overnight against 10 mM Tris/HCl, pH 8.2. The concentration of the nuclease was determined using an extinction coefficient of $\varepsilon_{280 \text{ nm}} = 44,620$ M/cm. The enzyme was stored in small aliquots at –80°C. After thawing, keep the nuclease at 4°C. For checking the activity of the nuclease and estimation how much enzyme is required for full digestion perform a methylation assay as described and digest the biotinylated oligonucleotides using different amounts of nuclease. The enzyme is commercially available as Benzonase from Merck.

8. Release of DNA by HCl was only checked in assays using MTases with good activity and was found to be equally efficient as enzymatic digestion. However, the influence of the HCl on the efficiency of Liquid Scintillation counting should be tested.

9. Mix the solution obtained after digestion and Liquid Scintillator Solution by vortexing or inverting the scintillation vials before counting.

References

1. A. Jeltsch (2002) Beyond Watson and Crick: DNA methylation and molecular enzymology of DNA methyltransferases. *Chembiochem* **3**, 274–93.

2. A. Hermann, H. Gowher, A. Jeltsch (2004) Biochemistry and biology of mammalian DNA methyltransferases. *Cell Mol Life Sci* **61**, 2571–87.

3. A. Bird (2002) DNA methylation patterns and epigenetic memory. *Genes Dev* **16**, 6–21.

4. P.A. Jones, D. Takai (2001) The role of DNA methylation in mammalian epigenetics. *Science* **293**, 1068–70.

5. R. Jaenisch, A. Bird (2003) Epigenetic regulation of gene expression: how the genome integrates intrinsic and environmental signals. *Nat Genet* 33 Suppl, 245–54.

6. P.A. Jones (2003) Epigenetics in carcinogenesis and cancer prevention. *Ann N Y Acad Sci* **983**, 213–9.

7. A.P. Feinberg, B. Tycko (2004) The history of cancer epigenetics. *Nat Rev Cancer* **4**, 143–53.

8. X. Cheng (1995) Structure and function of DNA methyltransferases. *Annu Rev Biophys Biomol Struct* **24**, 293–318.

9. E.J. Oakeley (1999) DNA methylation analysis: a review of current methodologies. *Pharmacol Ther* **84**, 389–400.

10. M.F. Fraga, M. Esteller (2002) DNA methylation: a profile of methods and applications. *Biotechniques* **33**, 632–49.

11. A.P. Bird, E.M. Southern (1978) Use of restriction enzymes to study eukaryotic DNA methylation: I. The methylation pattern in ribosomal DNA from Xenopus laevis. *J Mol Biol* **118**, 27–47.

12. A.P. Bird (1978) Use of restriction enzymes to study eukaryotic DNA methylation: II. The symmetry of methylated sites supports semi-conservative copying of the methylation pattern. *J Mol Biol* **118**, 49–60.

13. J. Singer, J. Roberts-Ems, A.D. Riggs (1979) Methylation of mouse liver DNA studied by means of the restriction enzymes msp I and hpa II. *Science* **203**, 1019–21.

14. M. Frommer, L.E. McDonald, D.S. Millar, C.M. Collis, F. Watt, G.W. Grigg, P.L. Molloy, C.L. Paul (1992) A genomic sequencing protocol that yields a positive display of 5- methylcytosine residues in individual DNA strands. *Proc Natl Acad Sci USA* **89**, 1827–31.

15. S.J. Clark, J. Harrison, C.L. Paul, M. Frommer (1994) High sensitivity mapping of methylated cytosines. *Nucleic Acids Res* **22**, 2990–7.

16. K.C. Kuo, R.A. McCune, C.W. Gehrke, R. Midgett, M. Ehrlich (1980) Quantitative reversed-phase high performance liquid chromatographic determination of major and modified deoxyribonucleosides in DNA. *Nucleic Acids Res* **8**, 4763–76.

17. C.W. Gehrke, R.A. McCune, M.A. Gama-Sosa, M. Ehrlich, K.C. Kuo (1984) Quantitative reversed-phase high-performance liquid chromatography of major and modified nucleosides in DNA. *J Chromatogr* **301**, 199–219.

18. D. Eick, H.J. Fritz, W. Doerfler (1983) Quantitative determination of 5-methylcytosine in DNA by reverse-phase high-performance liquid chromatography. *Anal Biochem* **135**, 165–71.

19. R.A. Rubin, P. Modrich (1977) EcoRI methylase. Physical and catalytic properties of the homogeneous enzyme. *J Biol Chem* **252**, 7265–72.

20. U. Hubscher, G. Pedrali-Noy, B. Knust-Kron, W. Doerfler, S. Spadari (1985) DNA methyltransferases: activity minigel analysis and determination with DNA covalently bound to a solid matrix. *Anal Biochem* **150**, 442–8.
21. A. Jeltsch, T. Friedrich, M. Roth (1998) Kinetics of methylation and binding of DNA by the EcoRV adenine-N6 methyltransferase. *J Mol Biol* **275**, 747–58.
22. M. Roth, A. Jeltsch (2000) Biotin-avidin microplate assay for the quantitative analysis of enzymatic methylation of DNA by DNA methyltransferases. *Biol Chem* **381**, 269–72.
23. P. Friedhoff, O. Gimadutdinow, T. Ruter, W. Wende, C. Urbanke, H. Thole, A. Pingoud (1994) A procedure for renaturation and purification of the extracellular Serratia marcescens nuclease from genetically engineered Escherichia coli. *Protein Expr Purif* **5**, 37–43.

14

Anti-Biotin Antibodies Offer Superior Organelle-Specific Labelling of Mitochondria Over Avidin or Streptavidin

Elisabeth D. Coene, Michael K. Shaw, and David J. Vaux

Summary

A number of endogenously biotinylated proteins are found in both cytosol and mitochondria of mammalian cells from many tissues, including liver, spleen, pancreas, kidney, and intestine. Therefore, caution should be taken when using the biotin detection system. Endogenous biotin can interfere with staining systems that employ the use of biotin–avidin- or biotin–streptavidin-based detection systems and may therefore result in high, non-specific background staining. Here, we show that this endogenous biotin reactivity can be deliberately exploited and used as a specific mitochondrial marker in both light and electron microscopy as well as for identifying mitochondrial fractions on Western blot.

Key Words: Biotin; avidin; streptavidin; electron microscopy; confocal laser scanning microscopy; antibodies; Western blotting; cryo-immuno EM; confocal immuno-fluorescence.

1. Introduction

The biotin carboxylase family is comprised of enzymes that utilize biotin as a co-factor. Biotin carboxylases are regulatory enzymes that play a role in diverse metabolic pathways such as lipogenesis and gluconeogenesis (1). They are widely distributed among prokaryotes and eukaryotes and are found in various tissues such as kidney, liver, salivary gland ducts, ductal epithelia (bronchial mucous glands), and some endocrine tissues (testis and adrenal)

From: *Methods in Molecular Biology, vol. 418: Avidin-Biotin Interactions, Methods and Applications*
Edited by: R. J. McMahon © Humana Press, Totowa, NJ

(2–5). While formalin fixation prevents detection of most endogenous biotin, it is still present in fairly high amounts. Moreover, antigen retrieval methods will unmask remaining endogenous biotin *(6)*. In some tissues, the endogenous biotin survives paraformaldehyde (PFA) fixation and will be revealed by Triton X-100 treatment *(7)*. Frozen sections of the above tissues too are likely to have endogenous biotin.

Endogenous biotin can interfere with staining systems that use the biotin–avidin/streptavidin detection method and can therefore result in high, non-specific background signal that can be interpreted as an erroneous positive result. While this non-specific background can be significantly reduced by pre-treatment of cells/tissues with avidin/biotin as blocking reagents prior to the incubation of biotinylated antibodies, we can exploit this phenomenon to obtain strong and highly specific mitochondrial labelling instead.

Here, we demonstrate that the presence of the endogenous biotinylated proteins can be used to obtain a strong and highly specific labelling method to detect mitochondria in cells, tissues, and subcellular fractions (*See* **Note 23**) and *(13)* using confocal laser scanning microscopy, ultra-thin freeze-thaw cryo-immuno electron microscopy (EM), and Western blotting, respectively.

2. Materials

2.1. Light Microscopy: Immuno-fluorescence Using Confocal Laser Scanning Microscopy

1. Human vascular endothelial ECV304 cells (The European Collection of Cell Cultures ref no. 92091712; Salisbury, Wiltshire, UK) are cultured in Dulbecco's Modified Eagle's Medium (DMEM) supplemented with 10% fetal calf serum (Invitrogen, CA, USA). They are grown at $37°C$, in a humidified incubator with 5% CO_2.

2. MitoTracker Red CMXRos (Molecular Probes, OR, USA) is a mitochondrial-selective stain that is concentrated by active mitochondria and well retained after fixation. Before opening a vial, allow the product to warm to room temperature (RT). To prepare a 1 mM stock solution, 50 μg of lyophilized MitoTracker is reconstituted in 94 μl of high-quality anhydrous dimethyl sulfoxide (DMSO) (Sigma-Aldrich, MO, USA). It can be stored frozen and protected from light for several months. Avoid repeated freezing and thawing. Dilute the 1 mM stock MitoTracker to the final concentration of 100–500 nM in prewarmed culture medium. Caution: DMSO is readily absorbed through the skin. Wear a lab coat and gloves when handling this solution. Skin, eye, and respiratory system irritant.

3. PFA (16%) (Sigma-Aldrich) stock: dissolve 16 g of PFA powder in about 80 ml dH$_2$O by stirring at approximately 55°C for 1 h (in fume cupboard). Do not exceed 60°C. Slowly add drops of 1 N NaOH (Sigma-Aldrich) to depolymerize

the PFA and keep on stirring (10 min after each drop) until the solution clears (keep at $\sim$60°C, be patient). Cool to RT and adjust the pH to 7.4. Make up to a final volume of 100 ml with dH$_2$O. Filter through a 0.45-µm Millipore filter (Millipore, Mass., USA). The 16% stock solution can be stored frozen in small aliquots and thawed when required (the thawed solution will again need warming to get all the PFA back into solution). Discard each thawed aliquot after use. Caution: PFA is very toxic, work in fumehood when preparing stock, do not inhale (wear mask), wear gloves; preparing the stock solution properly is essential for successful immuno-fluorescence.

   a. PFA fixative, 4%: dilute 16% PFA stock in appropriate buffer (1 M 4-(2-hydroxyethyl)-1-piperazineethanesulfonic acid (HEPES) buffer, pH 7.4; Invitrogen): 2.5 ml 16% PFA stock; 2.5 ml 1 M HEPES buffer; 5 ml dH$_2$O. Check the pH. Use immediately.

   b. PFA fixative, 8%: dilute 16% PFA stock in appropriate buffer (1 M HEPES buffer): 5 ml 16% PFA stock; 2.5 ml 1 M HEPES buffer; 2.5 ml dH$_2$O. Check the pH. Use immediately.

4. Phosphate-buffered saline (PBS): dissolve 1 PBS (Dulbecco A) tablet (Oxoid Limited, Hampshire, UK) in 100 ml dH$_2$O. Sterilize by autoclaving.

5. NH$_4$Cl, 50 mM: dissolve 0.134 g of NH$_4$Cl (Sigma-Aldrich) in 50 ml PBS buffer. Always prepare fresh.

6. Triton X-100, 0.5%: make from a 20% stock solution of Triton X-100 (Sigma-Aldrich). Caution: Viscosity increases as temperature falls and handling becomes difficult at temperatures below 20°C. Wear protective clothing and do not breath vapor. May be harmful if inhaled or in contact with skin. Causes severe eye irritation. The product may contain traces of ethylene oxide or dioxane, which are probable human carcinogens.

7. Cold water fish skin gelatin, 0.4%: warm gelatin from cold water fish skin (Sigma-Aldrich) in warm water bath at 37°C for 30 min to make it less viscous and easier to handle. Make a 4% stock solution of fish skin gelatin by mixing 4.4 ml of fish skin gelatin 45% solution with 45.6 ml PBS. We mix 5 ml of the 4% stock with 45 ml PBS to obtain 50 ml of 0.4 % fish skin gelatin solution.

8. Primary antibody: we used IgG fraction monoclonal mouse anti-biotin antibody (1.3 mg/ml) from Jackson ImmunoResearch Laboratories, Inc. (Cambridgshire, UK).

9. Secondary antibody: Alexa Fluor® 488-conjugated donkey anti-mouse IgG antibodies are used to detect the mouse monoclonal anti-biotin antibody. We coupled the Alexa Fluor® 488 label of Molecular Probes to a donkey anti-mouse secondary antibody from Jackson ImmunoResearch Laboratories, Inc., using the Molecular Probes Alexa Fluor® 488 Protein Labelling Kit.

10. Mowiol mounting medium, supplemented with 4′-6-diamidino-2-phenylindole (DAPI): make up 0.2 M Tris, pH 8.5 (MP Biomedicals, CA, USA), by dissolving 0.97 g Tris in 30 ml ultrapure (e.g., MilliQ) water. Adjust to pH 8.5 with 1 M

HCl (Sigma-Aldrich). Add MilliQ water to a final volume of 40 ml. Dissolve 2.4 g of Mowiol 4.88 Hoechst (Merck, Darmstadt, Germany) and 6 g of glycerol (Sigma-Aldrich) in 6 ml MilliQ water. Rotate the tube in a hybridization oven at 50°C overnight. Add 12 ml of 0.2 M Tris–HCl, pH 8.5, and rotate again in hybridization oven at 50°C for a few more hours. Centrifuge the solution at 600g for 5 min. Take off the supernatant and add DAPI (Sigma-Aldrich) to a final concentration of 0.1 µg/ml. Aliquot and freeze at –20°C protected from light.

11. We used the Radiance 2000 MP confocal laser scanning microscope from BioRad (BioRad Microscience, UK) equipped with the 900 Mira titanium-sapphire pulsed laser.

2.2. Electron Microscopy: Thawed-Frozen Ultra-Thin Section Immuno Electron Microscopy

1. Normal rat kidney tissue was obtained and processed as described below.
2. Four to eight percent PFA/250 mM HEPES, pH 7.4 (*see* **Subheading 2.1.**).
3. 20 mM, Glycine, in 250 mM HEPES buffer (pH 7.4): to make 250 mM HEPES buffer, mix 25 ml of 1 M HEPES buffer (Invitrogen) with 75 ml dH$_2$O. Dissolve 150 mg of glycine (Sigma-Aldrich) in 100 ml of 250 mM HEPES buffer. Check and adjust pH to 7.4.
4. 2.3 M, Sucrose, in PBS: dissolve 78.7 g of sucrose (Sigma-Aldrich) in 100 ml of PBS buffer.
5. 20 mM, Glycine, in PBS: dissolve 150 mg of glycine (Sigma-Aldrich) in 100 ml of PBS buffer.
6. 1%, BSA, in PBS: dissolve 1 g of BSA (Sigma-Aldrich) in 100 ml of PBS buffer.
7. Primary antibody: the same IgG fraction monoclonal mouse anti-biotin antibody (1.3 mg/ml) was used as for light microscopy.
8. Secondary antibody: we used a goat anti-mouse 10 nm gold-conjugated secondary antibody (British BioCell, UK).
9. Uranyl acetate (UA), 2%, in dH$_2$O (store in the dark at 4°C): dissolve 0.2 g of UA (Agar Scientific, UK) in 10 ml dH$_2$O. Uranyl ions react strongly with phosphate and amino groups so that nucleic acids and certain proteins are highly stained. Caution: UA contains trace amounts of U^{235} (radioactive). Powder should not be inhaled and solutions should be discarded in special containers.
10. 2%, Methyl cellulose (MC), in dH$_2$O: stir 2 g of MC (25 centipoise viscosity; Sigma-Aldrich) into 100 ml of cold dH$_2$O and leave overnight at 4°C. When fully dissolved, centrifuge at high speed (90–100,000 g) for several hours at 4°C. The MC should be removed without disturbing the pellet at the bottom of the tube and should be kept cold at all times. Caution: Avoid inhalation, contact with eyes, skin, and clothing. Avoid prolonged or repeated exposure.
11. Electron microscope: Tecnai 12 transmission electron microscope (FEI Company, The Netherlands) was used to analyze the samples.

3. Methods

3.1. Light Microscopy: Immuno-Fluorescence Using Confocal Laser Scanning Microscopy

1. ECV304 cells are grown on glass coverslips (25 mm diameter; thickness no. 1.5, VWR International Ltd (BDH), UK) in a six-well plate until a 70–80% confluency is obtained (*see* **Notes 1, 2**, and **3**).

2. Add 1–5 µl of 1 mM MitoTracker stock to 10 ml prewarmed DMEM culture medium. Mix well and add 2 ml of this Mitotracker solution to each well. Incubate for 10–15 min at RT. Remove the MitoTracker solution, wash the wells twice with prewarmed DMEM, and incubate the cells for another 10 min with prewarmed fresh DMEM culture medium.

3. Cell culture medium is removed and the coverslips are gently washed twice with ice cold PBS on ice.

4. Cells are fixed with 4% PFA/250 mM HEPES, pH 7.4, for 10 min on ice, followed by prolonged fixation in 8% PFA/250 mM HEPES, pH 7.4, for 50 min at RT. Fixative is removed and wells are filled with PBS.

5. Cells are quenched with 50 mM NH_4Cl for 10 min at RT (*see* **Note 4**) and subsequently washed with PBS for 10 min.

6. Cells are permeabilized with 0.5% Triton X-100 for 5 min at RT. The coverslips are thoroughly washed with four 10 min PBS washes at RT and changed to another six-well plate as the Triton X-100 detergent may still stick to the plastic even after intensive washing.

7. Block the non-specific binding sites on the coverslips with 0.4% cold water fish skin gelatin/PBS for 30 min at RT (*see* **Note 5**).

8. Incubate the coverslips with the primary antibody IgG fraction monoclonal mouse anti-biotin 1:200 diluted in 0.4% cold water fish skin gelatin/PBS for 1 h at RT (*see* **Notes 6** and **7**).

9. Wash 2 × 10 min in PBS at RT.

10. Incubate the coverslips with the secondary antibody donkey anti-mouse Alexa Fluor® 488 1:1000 diluted in 0.4% cold water fish skin gelatin/PBS for 1 h at RT. Cover the coverslips with a tray wrapped with aluminium foil to prevent unnecessary bleaching of the fluorescent label (*see* **Notes 6, 7**, and **8**).

11. Coverslips are washed extensively with PBS for 4 × 10 min at RT.

12. Prior to mounting, the coverslips are rinsed in dH_2O briefly to remove PBS traces, the excess dipped off on a tissue and mounted on glass slides in mounting medium with DAPI (*see* **Note 9**). Coverslips are stored in the dark in the fridge until they can be studied and imaged with a confocal microscope.

13. Example of the signals for both endogenous biotin and MitoTracker in mitochondria of ECV304 cells is shown in **Fig. 1**.

14. To determine if a signal is real and to differentiate it from any background labelling, we included a negative control by omitting the primary antibody (*see* **Note 10**).

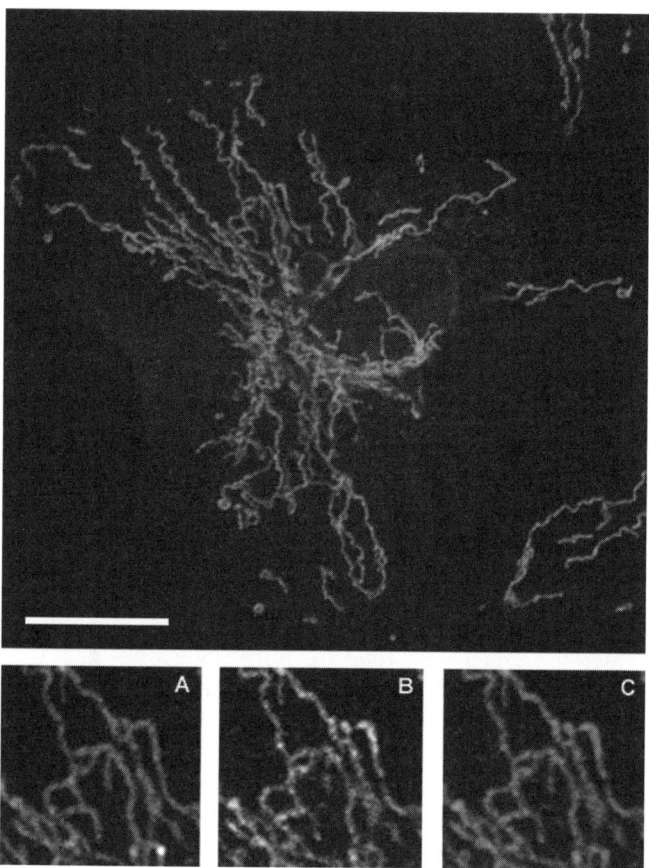

Fig. 1. Confocal microscopy image of ECV304 cell after immuno-fluorescence with MitoTracker Red CMXRos and anti-biotin antibody. ECV304 cell stained with anti-biotin antibody labelled with Alexa Fluor® 488 label (green), MitoTracker Red CMXRos (red). The nuclei are counterstained with 4´-6-diamidino-2-phenylindole (blue). Co-localization of mitochondria with endogenous biotin is shown, resulting in a yellow colour. Inserts are showing a higher magnification of (**A**) MitoTracker Red CMXRos, (**B**) anti-biotin Alexa Fluor® 488, and (**C**) the merged image. Bar: 10 µm.

3.2. Electron Microscopy: Thawed-Frozen Ultra-Thin Section Immuno Electron Microscopy

1. Pieces of tissue (in this case normal rat kidney) were fixed in 4% PFA in 250 mM HEPES buffer (pH 7.4) for 15–20 min at RT and then transferred into 8% PFA in 250 mM HEPES buffer (pH 7.4) for at least 30 min at 4°C (*see* **Note 11**).

2. After fixation, tissue samples are washed first with 250 mM HEPES buffer (pH 7.4). Tissue samples should then be washed with 250 mM HEPES buffer (pH 7.4) containing 20 mM glycine for 30–60 min at 4°C (*see* **Notes 12** and **13**).

3. Following these washes, the small pieces (~2 mm³) of tissue are transferred sequentially into increasing concentrations of sucrose (from 0.6 M up to 2.3 M sucrose) in PBS (or dH₂O as buffering is not critical at this stage) with several changes in the final sucrose solution. They are infused with 2.3 M sucrose overnight at 4°C although shorter times are possible (*see* **Note 14**).

4. Small (<0.5 mm³) pieces of cryo-protected/sucrose infiltrated tissue are then mounted on a copper or aluminium pin. The excess sucrose is removed with a filter paper and the pin is carefully plunged into liquid nitrogen. Specimens can be stored for years under liquid nitrogen (*see* **Note 15**). The process of cutting and retrieving ultra-thin cryo-sections is a relative simple technique to master although it does require some degree of skill and practice. The following sections are included for the sake of completeness although readers are encouraged to refer to either (*8*) or (*9*) for more detailed information on the technique.

5. Mount the specimen in the pre-cooled cryo-chamber of the cryo-ultramicrotome and cut the tissue block at a temperature of –100°C to obtain 65-nm-thin sections (*see* **Note 16**).

6. Pick up sections from the knife with a loop dipped in a 1:1 mixture of 2.3 M sucrose and 2% MC and transfer sections to a formvar/carbon-coated nickel grid. Grids are left floating section side down on PBS containing 20 mM glycine.

Immunogold labelling procedure for ultra-thin cryo-sections: the steps in immuno-labelling are outlined below. This method works for both thawed-frozen ultra-thin cryo-sections and sections of resin-embedded material. The process is conveniently carried out on a sheet of Parafilm with the grids floated, section side down, on droplets of solution and transferred from one droplet to the next using either a wire loop of approximately 5 mm diameter or with forceps. To ensure that the antibody droplets do not dry out (and concentrate), place a piece of wet filter paper near them and cover with a plastic dish. Routinely all media are made up in PBS (pH 7.4) although this is not critical and other buffers (e.g. Tris) can be used instead. All the incubations are performed at RT, except for the final contrasting and embedding step (*see* **step 13** below), which is done on ice (4°C).

7. Place the grids on separate drops of 1% BSA in PBS for 15 min. This step will block non-specific binding sites and hence lower any background labelling (*see* **Note 17**).

8. Using forceps lift the grids off the 1% BSA solution and remove the excess liquid using a filter paper. Place the grids on to drops (5–10 μl each) of the antibody solution. The antibody is diluted 1:100 using 1% BSA in PBS. Cover the grids and leave for 1 h at RT (*see* **Note 18**).

9. Rinse five times with PBS for at least 15 min.
10. Transfer the grids, using forceps and removing excess PBS, to a droplet of goat anti-mouse 10 nm gold-conjugated secondary antibody. The secondary antibody is diluted 1:50 using 1% BSA in PBS. Cover the grids and leave for 1 h at RT (*see* **Notes 18** and **19**).
11. Rinse five times with PBS for 30 min.

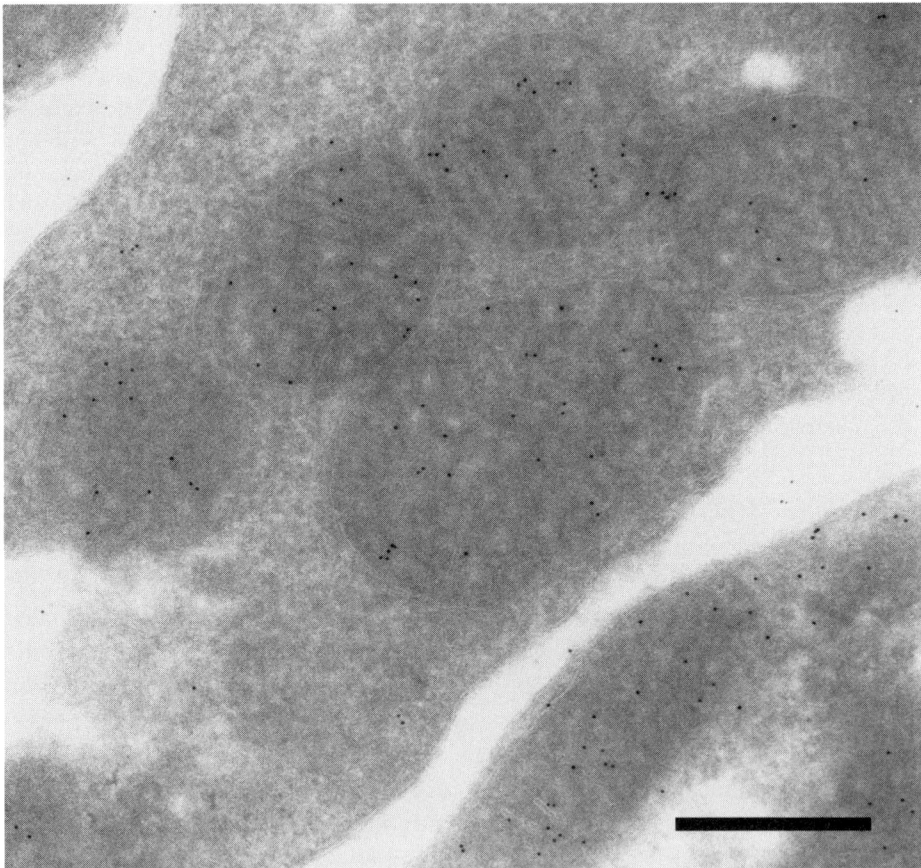

Fig. 2. (**A**) Electron microscopy (EM) image of normal rat kidney labelled for endogenous biotin. Section is labelled with a mouse monoclonal anti-biotin antibody and detected with a goat anti-mouse secondary antibody labelled with 10 nm gold. The signal is predominantly restricted to the mitochondrial matrix. Bar: 500 nm. (**B**) EM image of isolated mitochondria from rat liver labelled for endogenous biotin. Section is labelled with a mouse monoclonal anti-biotin antibody and detected with a goat anti-mouse secondary antibody labelled with 10 nm gold. The signal is predominantly restricted to the mitochondrial matrix. Bar: 100 nm.

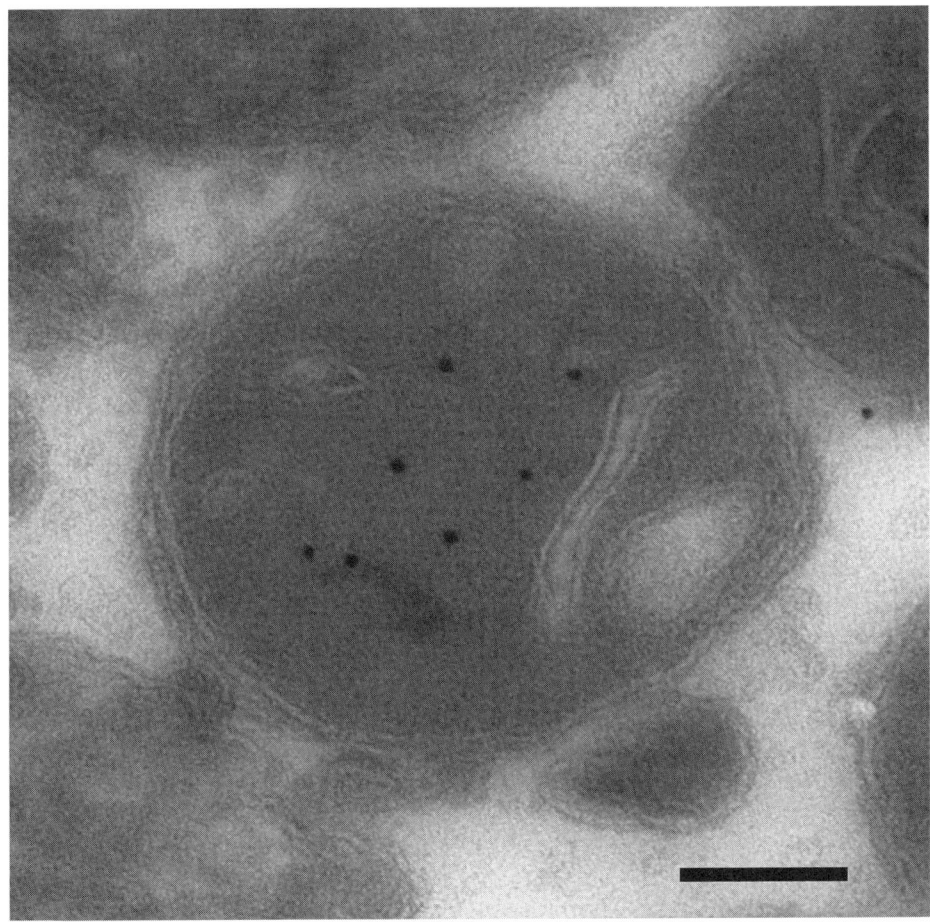

Fig. 2. *(Continued)*

12. Rinse several times with dH$_2$O for at least 20 min in total (*see* **Note 20**). The grids are now ready for contrasting and embedding.

13. UA staining and MC embedding: go quickly (seconds) over 2 drops of UA/MC (*see* **Note 21**), pH 4, and leave the section floating for 10 min on a third drop of that mixture. All UA/MC drops should be on ice. MC solutions are less viscous at lower temperature, so that they can penetrate better into the sections (*see* **Note 22**). Pick up with a 3.5-mm loop and remove excess liquid with a filter paper. Allow to dry before examining.

14. Tecnai 12 transmission electron microscope (FEI Company) was used to analyze the samples. The result of the immuno EM using the mouse monoclonal anti-biotin primary antibody and detected with the goat anti-mouse 10 nm gold-conjugated secondary antibody on normal rat kidney tissue is shown in **Fig. 2A**.

15. To determine if a signal is real and to differentiate it from any background labelling, we included a negative control by omitting the primary antibody (*see* **Note 10**).

4. Notes

1. Coverslips can be obtained in different sizes and thicknesses. Keep in mind that each microscope's objective requires a specific coverslip thickness for optimal results. The coverslips are autoclaved or can be flamed in EtOH to render them sterile before putting them in the six-well plates.
2. Never let the staining areas dry out. This will increase background staining tremendously.
3. This method is also applicable for many other cell lines (e.g., NRK and HeLa cells) or can be performed on tissue sections. To obtain ultra-thin tissue cryo-sections we follow the same procedure for EM as described in **Subheading 3.2**.
4. Quenching: This solution is used to quench the free aldehyde groups that are created when using PFA fixation and might bind your antibodies non-specifically. Always prepare fresh immediately prior to use.
5. After the tissue has been incubated with the blocking buffer, do not rinse off the blocking agent, simply drain the excess.
6. Dilute primary and secondary antibody in blocking solution and centrifuge the solution at maximum speed for 1 min in an Eppendorf centrifuge to pellet protein aggregates. To ensure that the antibody droplets do not dry out (and concentrate), place a piece of wet filter paper near them and cover with a plastic dish.
7. To spare precious antibody, the coverslips can be incubated on a 100-µl drop on Parafilm. To lift the coverslip without damaging the cells, pipette PBS underneath the coverslip first before lifting it with forceps. The coverslips are put back, cells facing up into the six-well plate.
8. If a fluorescent marker is being used, check to make sure that there is no auto-fluorescence in the unprocessed, fixed tissue or cells. If there is auto-fluorescence, choose a fluorescent marker that will not compete. From this step on, it is preferred to work in the dark or cover the coverslips with a tray wrapped with aluminium foil to prevent bleaching.
9. When mounted in Mowiol (small drop) let polymerize for 1 h at RT and store the slides at 4°C in the dark (aluminium foil or slide maps).
10. To determine if a signal is real and to differentiate it from any background labelling, both negative and positive controls must be included in any labelling protocol. Negative controls should include one or all of the following: omit the primary antibody; use a non-specific primary antibody of the same species; absorb the primary antibody with antigen before incubation; use a non-specific second antibody. These negative controls will help identify the source of any non-specific background. A good positive control using a specimen with high antigen content will test the whole labelling system.

11. Fixation with low PFA concentrations (<4%) is, to a considerable extent, reversible *(10,11)*. Thus extensive washing of the specimen after PFA fixation should be avoided if possible. The situation is less critical for higher concentrations of PFA (8%) as the cross-linking process is more efficient.

    A similar fixation protocol can be used for both cultured cells and subcellular fractions but embedding in gelatin might be advisable (*see* **Note 13**). We included an EM image of a mitochondrial fraction, prepared from rat liver, labelled for endogenous biotin using the above-mentioned method in combination with **Note 13** (*see* **Fig. 2B**).

12. Glycine binds to free aldehyde groups from the fixative that may still be available for cross-linking antibodies and lead to background labelling. Ammonium chloride (50 mM) or lysine (20 mM) may be used instead of glycine.

13. For some samples (e.g., loose tissues such as lung, spleen, or pancreas), cultured cells and cell fractions, it is advisable to embed them in gelatin after fixation to create a more cohesive block that is easier to handle, more convenient for trimming, and is much easier to section. For further details *see* **ref. 9**.

14. The use of sucrose as a cryo-protectant was introduced by Tokuyasu *(12)* and is still the best method for freezing and sectioning specimens. The optimal sucrose concentration(s) depends on the tissue and the fixative used. Sucrose, 2.3 M, in PBS is routinely used in our lab to infuse tissues and cell pellets. Infusion times range from 30 min to overnight depending on the type and size of the specimen. One way to determine if a specimen is sufficiently infused with sucrose is to push the blocks down into the sucrose and watch if they quickly float back to the top of the sucrose. If they stay down then they are properly infused, if they float back to the surface then they require a longer incubation.

15. Remove excess sucrose from the surface of the specimen before freezing—it is time consuming to cut away large amounts of sucrose, and because the frozen block is uniformly white, it may cause you to cut and collect only sections of sucrose. Avoid air bubbles between the specimen and the holder; the block may fall off the holder while it is being frozen. Do not let the specimen dry out when manipulating it on the holder.

16. It is important to select the correct temperature for cutting as section thickness is dependent upon temperature for tissue infused with a given sucrose concentration. The higher the sucrose concentration the softer the tissue block, and a colder cutting temperature is required for any given section thickness. As a general rule, the thinner the section required, the colder the temperature required. When sectioning sucrose-embedded material, the aim should be to cut sections as thin as possible. Frozen sections show interference colours that are related to their thickness. The actual thickness of sections with gold interference colours is around 110 nm.

17. Non-specific binding sites on the sections should be blocked using a 'sticky' protein. We use 1% BSA in PBS but normal (i.e., non-immune) serum of the secondary antibody species, 1% gelatin (calf skin or cold water fish skin), 1% ovalbumin, or 5–10% fetal bovine serum can also be used as alternative blocking

reagents. The normal serum should be heat-inactivated and filtered (0.2 μm) before use. Another method for reducing background labelling (reducing positive charges) is to raise the pH of the incubation buffer.

18. Both primary and secondary antibody, diluted to a suitable concentration, in PBS containing blocking agent, should be centrifuged in a bench-top microfuge for 1–2 min before use. This will remove any antibody aggregates formed during storage.

19. The usual size range of gold probes for TEM is 5–20 nm. Using smaller gold probes (5–10 nm) has the advantage of improved spatial resolution although 5-nm gold particles may be hard to visualize over densely stained structures. With the larger gold probes, effects of steric hindrance may also decrease the labelling density.

20. This step is very important. Do not forget it. Salt solutions are used for diluting the antibodies and for washes. If the PBS is not completely washed away, then any phosphate ions present will precipitate the uranyl salts onto the sections during the contrasting and embedding stage. Therefore, wash specimens with water before incubation with UA.

21. The standard mixture is made by mixing 9 parts of 2% MC with 1 part of 2% aqueous UA. But other concentrations can by tried as well *(9)*. Higher concentrations of UA result in better negative staining of membranes but also cause a general increase in electron density, which makes the gold particles more difficult to see.

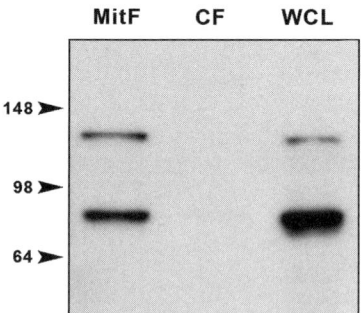

Fig. 3. Western blot of mitochondrial (MitF), cytosolic (CF), and whole cell lysate (WCL) HeLa fractions. Endogenous biotinylated proteins are detected using anti-biotin antibody and donkey anti-mouse horseradish peroxidase secondary antibody. Arrowheads mark molecular weight standards at 64, 98 and 148 kDa. Two major species of 75 kDa and 130 kDa are detected in both the mitochondrial and whole cell lysate fractions whereas the cytosolic fraction is clean. A much weaker signal may sometimes be seen in the cytosolic fraction on higher loading or longer exposure; this corresponds to a low abundance endogenously biotinylated cytosolic protein, acetyl-CoA carboxylase, and is easily distinguished from the mitochondrial carboxylases by its distinctive molecular weight (270 kDa).

22. If MC is used to support cryo-sections during drying, a slightly opaque, globular precipitate sometimes appears. This may be due to either the MC being (i) uncentrifuged, (ii) stored too long, or (iii) warmed prior to use. Make a fresh solution of MC and centrifuge it prior to use.

23. The presence of endogenous biotin has also been shown to be a valuable and reliable tool to identify subcellular mitochondrial fractions on Western blot (*see* **Fig. 3**). A description of the fractionation and blotting methods is given in detail in **ref.** *13*.

5. Acknowledgments

E.D. Coene is a Postdoctoral Researcher employed by the Fund of Scientific Research Flanders. D.J. Vaux would like to acknowledge research support from the Wellcome Trust, the Medical Research Council, the EP Abraham Trust, and Synaptica Ltd.

References

1. Wolf, B. (1995) Disorders of biotin metabolism, in *Metabolic and Molecular Basis of Inherited Disease* (Scriver C.R., Beaudet, A.L., Sly, W.S., Valle, D., eds) 7th ed. New York, McGraw-Hill, pp 3151–77.

2. Wood, G.S., and Warnke, R. (1981) Suppression of endogenous avidin-binding activity in tissues and its relevance to biotin-avidin detection systems. *J. Histochem. Cytochem.* **29**, 1196–204.

3. Dakshinamurti, K., and Mistry, S.P. (1963) Tissue and intracellular distribution of biotin-C-1400H in rats and chicks. *J. Biol. Chem.* **238**, 294–6.

4. Green, M., Sviland, L., Taylor, C.E., Peiris, M., McCarthy, A.L., Pearson, A.D., and Malcolm, A.J. (1992) Human herpes virus 6 and endogenous biotin in salivary glands. *J. Clin. Pathol.* **45**, 788–90.

5. Kirkeby, S., Moe, D., Bog-Hansen, T.C., and van Noorden, C.J. (1993) Biotin carboxylases in mitochondria and the cytosol from skeletal and cardiac muscle as detected by avidin binding. *Histochemistry* **100**, 415–21.

6. Bussolati, G., Gugliotta, P., Volante, M., Pace, M., and Papotti, M. (1997) Retrieved endogenous biotin: a novel marker and a potential pitfall in diagnostic immunohistochemistry. *Histopathology* **31**, 400–7.

7. Satoh, S., Tatsumi, H., Suzuki, K., and Taniguchi, N. (1992) Distribution of manganese superoxide dismutase in rat stomach: application of Triton X-100 and suppression of endogenous streptavidin binding activity. *J. Histochem. Cytochem.* **40**, 1157–63.

8. Griffiths, G. (1993) *Fine Structure Immunocytochemistry*. Springer-Verlag, Heidelberg.

9. Tokuyasu, K.T. (1997) Immuno-cytochemistry on ultrathin cryosections, in *Cells, A Laboratory Manual. Vol. 3 Subcellular Localization of Genes and Their*

Products. (Spector, D.L., Goodman, R.D., and Leinwand, L.A., eds.). Cold Spring Harbor, Cold Spring Harbor Laboratory Press, pp 131.1–.27.

10. Tokuyasu, K.T., and Singer, J.S. (1976) Improved procedures for immunoferritin labelling of ultrathin frozen sections. *J. Cell Biol.* **71**, 894–906.

11. Tokuysau, K.T., Dutton, A.H., and Singer, S.J. (1981) Ultrastructure of chicken cardiac muscle as studied by double immunolabelling in electron microscopy. *Proc. Natl. Acad. Sci. U. S. A.* **78**, 7619–23.

12. Tokuyasu, K.T. (1973) A technique for ultracryotomy of cell suspensions and tissues. *J. Cell Biol.* **57**, 551–65.

13. Coene, E.D., Hollinshead, M.S., Waeytens, A.A., Schelfhout, V.R., Eechaute, W.P., Shaw, M.K., Van Oostveldt PM, and Vaux DJ. (2004) Phosphorylated BRCA1 is predominantly located in the nucleus and mitochondria. *Mol. Biol. Cell* **16**, 997–1010. Epub Dec 9, 2004.

15

Mammalian Lectin as Tool in Glycochemistry and Histochemistry With Relevance for Diagnostic Procedure

Karel Smetana, Jr., and Sabine André

Summary

Carbohydrates represent a fundamental building unit of living organisms. Many contemporary results introduce these substances as medium with remarkable data storage capacity (glycocode) that is decoded by special receptor counterpartners, lectins. Animal so-called endogenous lectins are presented here as the biotinylated tools for normal lectin histochemistry in accompany with biotinylated (neo)glycoconjugates employed in reverse lectin histochemistry. Practical lesson how to employ these probes in cell/tissue labeling including multiple cell labeling at the single-cell level is also included. Position of glycocode and endogenous lectins in squamous epithelium biology under physiological condition and after the malignant transformation is shown as example of the employment of mentioned probes in research and diagnostics.

Key Words: Carbohydrate; lectin; galectin; (neo)glycoconjugates; lectin histochemistry; reverse lectin histochemistry; squamous cell epithelium; squamous cell carcinoma.

1. Carbohydrate Chain Variability, Glycocode

Saccharides were, according to older textbooks of biochemistry and cell biology, considered predominantly for a source of energy and/or for important structural components, for example, of extracellular matrices or cell surface. They were described as substances with high ability to bind water because of the high content of hydroxyl and other hydrophilic functional chemical groups. This phenomenon influences physicochemical properties of majority of living tissues. Compare for example extracellular matrix of hyaline cartilage and the

From: *Methods in Molecular Biology, vol. 418: Avidin-Biotin Interactions, Methods and Applications*
Edited by: R. J. McMahon © Humana Press, Totowa, NJ

humor vitreous from the eye. Last decade is connected in the literature with bioinformatics role of these interesting substances. Monosaccharides have a remarkable ability to polymerize to variable linear or branched chains. The number of linear isomers (LI) from the defined number of monosaccharides can be calculated according formula *(1)*

$$LI = E^n \times 2^n \times 4^{n-1}, \qquad (1)$$

where E^n represents the permutations from order of sequences including repetitions of the same sugar. Each hexose can be in pyranose or furanose form (2^n) and it can occur in two anomeric forms (α/β) (2^n). The linkage between two hexoses can be at four potential hydroxyls (4^{n-1}).

If we will calculate the number of linear hexasaccharides prepared from six hexoses according to this formula, we will receive 195,689,447,424 linear isomers. Although a majority of them is theoretical only and they are not occurred in nature, this number is remarkable if it is compared with number of hexapeptides prepared from six amino acids (46,656) *(1)*. When we contemplate that majority of saccharides is branched (the formula and calculation are not shown), the number of isomers will be higher. On basis of these theoretical findings, we can speculate that carbohydrates represent almost ideal medium for storage of biological information, and the term glycocode (sugar code) was introduced to the contemporary cell and molecular biology *(2–5)*.

Except for not so frequent free saccharidic chains, majority of saccharides exists in form of glycoconjugates. Except the nucleic acids that are out of topic of this article, they exist mainly in form of glycoproteins and glycolipids. Peptides/proteins can be glycosylated in distinct epitopes such as serine/threonine, where saccharidic chain is attached via hydroxyl group by so-called *O*-glycosidic bond. The second site that can be glycosylated is the asparagine through the amidic group of amino acid, so-called *N*-glycosidic bond. The glycosylation represents the most important posttranslational modification of proteins and strongly influences their structure and functional properties *(6)*. Glycolipids are glycoconjugates where carbohydrate is attached by esteric bond to the primary hydroxyl group of ceramide. These substances are usually present in cell membranes and they influence their functional properties, namely, they are responsible for their hydrophilicity *(6)*. However, the bioinformatics role of sugar motifs as a component of glycoproteins/lipids must be also reasoned.

2. Mammalian Lectins

2.1. History of Lectins, from Invertebrate to Plant Lectins

When will suppose that epigenetic biological information is stored in sugar oligo/polymer chain, this information must be decoded by specific manner.

Distinct sugar motifs are specifically recognized by antibodies. It is known that natural antibodies physiologically occur in serum, which specifically recognize ABO histo-blood saccharidic antigens and that they are very important for successful blood transfusion. Some enzymes (glycosyltransferases and glycosy-dases) also recognize distinct saccharidic motifs *(7)*.

All living organisms from bacteria to humans contain proteins/glycoproteins different from the above-mentioned antibodies and enzymes that also recognize distinct saccharidic motifs. They are called lectins (from Latin word *legere*, which means to select or choose). According to Roger A. Laine *(1)*, seven structural motifs are important for precise recognition on carbohydrate by lectin. They include epimers (including *D* and *L*), sequence of sugars, anomeric configuration, ring size, linkage position, branching and charge (for example, carboxylate, sulfate or amino group).

History of lectin research is 145 years old (the lectin research history is dealt in detail in **(5)** and **(8)**. The coagulation of blood by rattlesnake venom based on lectin mechanism was observed in 1860, and the first lectin was described from highly toxic castor bean seeds by Stillmark in 1888. This type of lectin was employed by Paul Ehrlich in research of neutralizing antibodies in 1891. Bacterial agglutinins were discovered by Kraus in 1902. The last century is connected with explosive description of plant and invertebrates lectins. Role of lectins occurred in viruses and bacterial cells is connected predominantly with their virulence, and plant lectins participate predominantly in defense against pathogens and protection from herbivorous animals *(5,9)*. However, these lectins mainly represent the exciting tool for the specific detection of distinct saccharidic motifs in the hands of biologist, and they are widely employed in cell biology and histochemistry. The extensive research of animal, so-called endogenous, lectins is connected with last two decades of previous century and it continues also in these days.

2.2. Classification and Function of Mammalian Lectins

Animal lectins represent structurally heterogeneous group of proteins and glycoproteins, and according to structural homology, five classes of animal endogenous lectins were discovered (*see* **Table 1**). For detailed description refer the works by Gabius *(10)* and Kaltner and Gabius *(11)*. Their function is heterogeneous, but their ability to recognize distinct saccharidic motif is in the center of their activity. As summarized in **Table 1**, it includes processes involved in the immune recognition, predominantly included to the innate immunity. However, their role as adhesive molecules in the process of intercellular and cell-extracellular matrix interaction is also remarkable *(12)*. Last but not least is participation of endogenous lectins in the control of cell cycle and apoptosis *(13)*. The extensive research in endogenous lectins and their

Table 1
Classification of Animal Lectins According to Structural Motifs

Lectin family	Number of known lectins	Example	Function
C-type	>40	Selectins	Migration of leukocytes through endothelium
	175-kDa mannose receptor	Antigen recognition and presentation	
Galectins	15	Galectin-1	Proapoptotic, cell–cell interaction
	Galectin-3	Antiapoptotic, cell–cell and cell–substratum interaction, macrophage activation	
I-type	>10	Siglecs	Cell–cell interaction in immune and nervous system
P-type	2	46-kDa mannose-6-phospahte-binding protein	Targeting of lysosomal enzymes
Pentraxins	2	C-reactive protein	Innate immune protection

glycoligands located in cancer cells was performed, and these results brought highly promising results concerning the study of tumor formation, diagnosis and selection of new prognostic markers and selective drug targeting to cancer cell (14–17).

2.3. Carbohydrate Recognition Domain

Carbohydrate recognition domain (CRD) represents the part of lectin molecule, which is responsible for the specific recognition of carbohydrate motif by lectin. Kurt Drickamer proposed (18) that only limited part of lectin is able to recognize sugar moiety, and he started to use the term CRD.

Several conservative types of CRDs were discovered and they are one of the most important structural features for classification of endogenous lectins. CRD recognizes the sugar counterpartner by noncovalent interaction based on hydrophobic and electrostatic interaction. Crystallographic and thermodynamic studies and site mutagenesis experiments were performed to characterize the most important amino acids and their position in CRD of distinct classes of animal lectins *(19,20)*.

The binding specificity of endogenous lectin members of one class can be very similar, but they are fine specificities concerning the saccharide binding. For example, family of galectins includes 15 members. Galectins can be divided into three subfamilies. (i) The so-called prototype galectins contain one CRD, which can form bivalent dimmers. (ii) Chimera-type galectins also contain one CRD and long N terminus, with which galectins can form multivalent oligomers *(21)*. (iii) Tandem-repeat galectins are composed of two CRDs connected with a short chain (*see* **Table 2**). Although CRD is conservative in all members and galectins recognize β-galactosides, they are fine specificities in CRD reactivity in different members of galectin family *(22)*.

In addition to carbohydrates, lectins also recognize protein ligands *(23)*. For example, galectin-1 and galectin-3 interact directly with protein Gemin4, one of the components of snRNPs participating in pre-mRNA splicing *(24)*. Galectin-3 also interacts with β-catenin, where it involves Wnt/β-catenin signaling pathway *(25)*.

Proteins different from antibodies and enzymes and with no CRD can also specifically recognize defined saccharidic moieties. For example, heparin is selectively recognized by histones in contrary to other anionic polysaccharides. However, histones lacking CRD cannot be considered for lectins according to classical definition *(26)*.

Table 2
Types of Galectins

Type of galectin	Structure	Members
Prototype		1, 2, 5, 7, 10, 11, 13, 14, 15
Chimera		3
Tandem repeat		4, 6, 8, 9, 12

, carbohydrate recognition domain; , N-terminus; , Connection between 2 carbohydrate recognition domains.
Prepared according to Liu and Rabinovich *(17)*.

3. Tool for Study of Mammalian Lectins, Their Glycoligands and Binding Reactivity

Biologist and histologists have powerful tool in their hands to decode the glycocode in situ by functional lectin and glycome analysis. Expression of endogenous lectins, detection of their reactive glycoligands and binding reactivity of these lectins can be monitored in cells and tissue sections.

3.1. Detection of Mammalian Lectins

Mammalian (endogenous) lectins can be detected by conventional immuno-histochemistry. Because, some lectins have conservative structure (for example, members of galectin family), it is necessary to be very careful concerning the cross-reactivity of antibody inside the galectin family. Control of antibody specificity using western blot analysis is necessary, namely, in case of polyclonal antibody.

3.2. Detection of Glycoligands

Distinct saccharidic motif can be detected using immunohistochemistry by specific antibodies. However, the biotinylated lectins represent the most frequently used probe for specific detection of saccharides. The lectins of plant and invertebrate origin are routinely used. The isolation and production of recombinant human lectins opened new horizons in the detection of distinct saccharide motifs. Because these lectins naturally occur in tissues, the visualization of their saccharidic counterpartners has a high physiological relevance. Moreover, the size and three-dimensional structure of lectins can influence the availability of glycoligand. Comparing the plant lectin and animal lectin with the similar reactivity, the binding of both lectins to tissue can substantially differ *(27)*, which can be influenced by the difference of the accessibility of glycoligand for natural and exogenous receptor. The biotinylation of endogenous lectins is the mostly frequently used procedure, how to label these lectins for further employment as probes *(28)*. Biotinylation procedure can influence the properties of lectin such as shift of pI. It can influence the unspecific binding of lectin *(29)*.

3.3. Detection of Binding Reactivity of Animal Lectins Present in Cells and Tissues

Binding reactivity of endogenous lectins that occur in cells and tissues can be studied by semisynthetic/synthetic (neo) glycoconjugates. They usually represent the conjugate of specific saccharidic motif with macromolecule, because it is known that carbohydrates attached to the polymer (albumin or polyacrylamide chains) are more reactive for endogenous lectins than the saccharidic motif only *(30,31)*. The use of multivalent synthetic saccharidic ligands based on dendrimers seems to be also advantageous for the detection of carbohydrate-binding sites *(32)*. These substances are also prepared in biotinylated form for use in histochemistry *(33)*.

3.4. How to Use Biotinylated Mammalian Lectins and (Neo)glycoligands as Probe in Single/Multiple Labeling at Single-Cell Level Instruction Manual

Three types of probe, that is, antibodies, biotinylated lectins and biotinylated (neo)glycoligands, represent a powerful arsenal for multiple labeling at single-cell level. It is necessary to mention here that frozen tissue is more suitable for analysis than paraffin embedded, because the tissue treatment for paraffin embedding is connected with removal of lipid substances and deprivation of glycolipids. From this point of view, the fixation of cells and tissue sections by paraformaldehyde is more suitable than use of methanol or acetone, fixatives commonly used in immunohistochemistry. These methodic principles are described in detail in **ref. 34**.

3.4.1. Detection of Endogenous Lectin Expression and Binding Reactivity of These Lectins (Reverse Lectin Histochemistry)

Concerning the strategy, it is necessary to use monoclonal antibody, which does not influence recognition of carbohydrate motifs. However, the expression of endogenous lectin and carbohydrate-binding site at the same cell level cannot be considered for evidence, that this concrete lectin recognizes the used motif. This observation must be verified by inhibition of carbohydrate binding by monoclonal antibody blocking the sugar binding to CRD of studied lectins (*see* **Fig. 1**) *(35)*. Describing the concrete method, after fixation and extensive washing of specimen by PBS, the nonspecific adsorption of antibody and/or biotinylated (neo)glycoligand must be prevented. Similarly to conventional immunohistochemistry, the diluted nonimmune serum of animal species from which the second step antibody was prepared is suitable. This blocking solution also prevents the nonspecific binding of antibodies via Fc receptors. If the blocking with nonimmune serum is not possible, for example, because it interacts with (neo)glycoligand, it can be replaced by incubation with albumin. The first step antibody against endogenous lectin and biotinylated (neo)glycoligand can be applied simultaneously at one step. After the washing with PBS, the labeled second step antibody and avidin-based substance are applied. We recommend the second step reagents labeled by fluorochroms. The classical combination FITC and

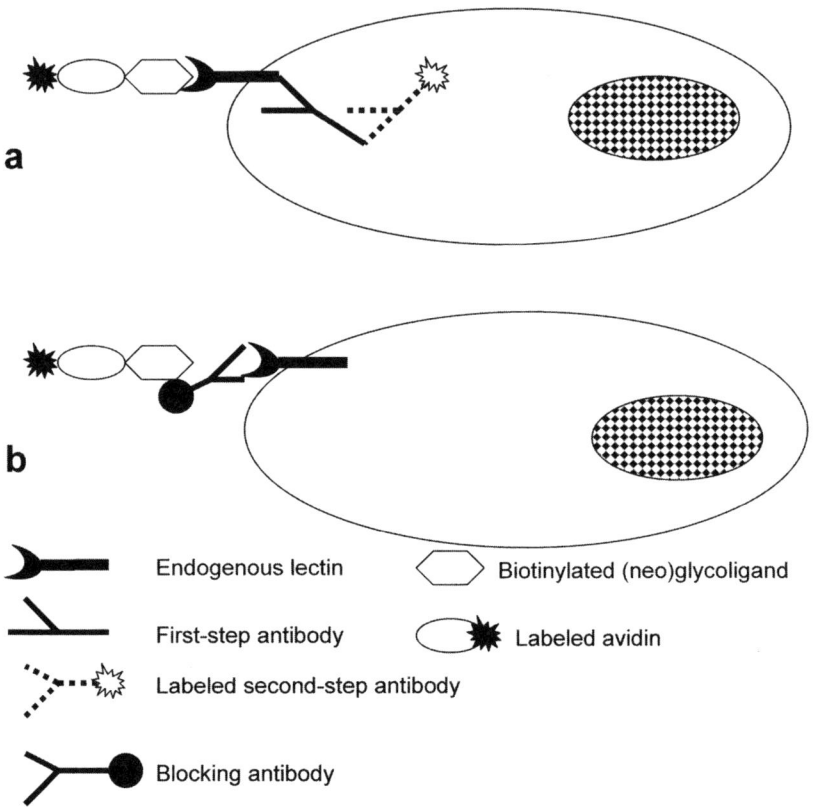

Fig. 1. Binding of labeled (neo)glycoligand to endogenous lectin identified by immunocytochemistry (a) is blocked by antibody recognizing the carbohydrate recognition domain of this lectin (b). This approach is necessary to verify binding reactivity of lectin for distinct sugar motif *in situ.*

TRITC is very convenient. The counterstaining of nuclei with DAPI or Hoechst is important for easy orientation in the specimen. The use of fluorochroms is advantageous, because it is possible to monitor each marker separately using specific filter blocks. The second step reagents (antibody, avidin) can be also labeled by enzymes, for example, alkaline phosphatase and peroxidase. The reaction products are dark brown/blue, and in case of intensive reaction, it is not so easy to distinguish both positivities. Similarly to other histochemical reaction, the control of the reaction specificity is an important step for histochemical reaction. The specificity of antibody binding can be tested by the replacement of specific antibody by the antibody of the same isotype against antigen not occurring in the specimen. The blocking of neoglycoligands binding by saccharide in the incubation solution is a good control of the specificity of carbohydrate-binding site reactivity. Because avidin-based substance can also recognize naturally occurred ligands in human cells and tissues, it is necessary to test the binding of avidin to the specimen.

3.4.2. Detection of Expression of Endogenous Lectin and Carbohydrate Reactive for This Lectin at Single-Cell Level

This methodological approach is in comparison with previously described, rather complicated methods, because both studies used probes (antibody against lectin and biotinylated lectin) that mutually interact. The possible way out from this problem is in the employment of modified lectin if it is available. For example, galectin-3 has long N terminus (*see* **Table 2**). When the antibody against N terminus is employed simultaneously with truncated form of galectin without -terminus *(36)*, both activities, that is, galectin-3 expression and binding sites for galectin-3 at single-cell level, can be detected simultaneously at the single-cell level (*see* **Fig. 2**).

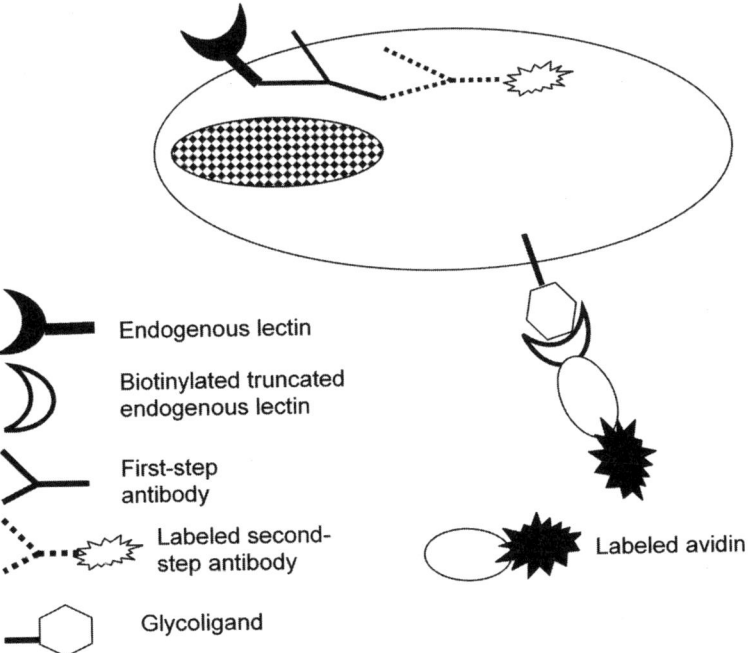

Fig. 2. Example of simultaneous detection of expression of endogenous lectin and ligands for this lectin at single-cell level by combination of immunocytochemical detection of galectin-3 by monoclonal antibody against the N- terminus of galectin and lectin histochemistry using biotinylated truncated galectin-3 lacking the N-domain.

3.4.3. Simultaneous Detection of Expression of Carbohydrate-Binding Sites or Lectin-Reactive Ligands with Reference Markers

Carbohydrate binding sites or endogenous lectin-reactive ligand expression monitored by biotinylated probes (biotinylated (neo)glycoprotein or lectin) can be detected simultaneously with reference marker, which expression is typical for some functional status of studied cell population. We can, for example, study binding sites for galectin-3 in company with expression of cytokeratins, which enables to estimate the differentiation status of studied cells or β1 integrin located in proliferative cells of epithelium. The combination of immunocytochemistry with normal or reversed lectin histochemistry is highly suitable for this purpose *(34)*.

4. Employment of Labeled Endogenous Lectins and Neoglycoligands in Biology and Histopathology Demonstrated in Squamous Cell Epithelia and Related Cancer

Squamous cell epithelia cover body surface (epidermis) and cavities (oral mucosa, laryngeal epithelium and esophagus). It represents functionally and morphologically stratified tissue with compartment of proliferating cells in basal layer. Some of these cells have properties of stem cells. Other stem cells, especially, of epidermis are located in bulge region, outer root sheath of hair follicle *(37)*. Stem cells of hair follicle exhibiting special pattern of keratins (K15 and K19), cytoplasm and ΔNp63α in the nucleus express galectin-1 and binding sites reactive for this galectin in their nuclei *(29,38,39)* in sections of human and porcine skin as well as in cultured cells. Performing the colocalization with splicing factor SC35, galectin-1-binding sites are localized in nuclear speckles indicating some role in splicing of pre-mRNA *(29)*. Cytoplasm of basal proliferating and suprabasal postproliferative cells as well as all cells from carcinomas of the skin and head and neck is reactive for galectin-1 *(40)*. In contrary to these results, galectin-3, galectin-3-reactive glycoligands and binding sites for galectin-3-reactive biotinylated asialofetuin were observed suprabasally in normal squamous epithelia and in the most differentiated portions of squamous cell carcinomas *(40–43)*. Interestingly, these galectin-3-reactive ligands are present predominantly in the cell surface and colocalize with desmosomal proteins *(42)*. Basal absence of accessible galectin-3-reactive glycoligands seems to be connected with expression of α2,6-linked *N*-acetylneuraminic acid, because removal of this monosaccharide restored galectin-3 reactivity of basal cells *(44)*. The head and neck squamous cell carcinomas with cells exhibiting binding sites for galectin-3 form lymph regional metastases in lower extent than tumors with low level of galectin-3-reactive ligands, and prognosis of these patients is significantly better

(45). Galectin-3-reactive glycoligand expression can be considered for new independent prognostic factor of head and neck carcinomas *(46)*. Galectin-7 expression is specific for squamous cell type epithelium, where it is related to proapoptotic p53 activity *(47,48)*. The carcinomas originated from squamous cell epithelium exhibited down-regulation of galectin-7 expression *(49)*. Gene transfection-induced overexpression of galectin-7 inhibited growth of squamous cell carcinoma cells *(50)*.

Dendritic cells as professional antigen presenting cells of Langerhans type occur in squamous cell epithelia, where they play a crucial role in protection of organism against microbial invasion and cancer formation *(51)*. These cells exhibit binding sites for galectin-3, and this lectin is internalized into the Birbeck granules-specific organelle important in the process of antigen presentation *(52,53)*. Other endogenous lectin, mannosides-reactive Langerin controls formation of Birbeck granules *(35,54)*.

The results of the study of endogenous lectins, their reactive glycoconjugates and carbohydrate-binding sites in squamous cell epithelia demonstrate the importance of glycoconjugates and lectin in control of epithelial biology under physiological and pathological condition including the immune surveillance.

5. Conclusion

Study of the expression of endogenous lectins, reactive glycoconjugates and carbohydrate-binding sites represents a perspective field of cell biology, histology and histopathology with possible clinical implications. Biotinylated probes such as endogenous lectins and (neo)glycoconjugates are in the center of this effort.

Acknowledgments

Part of results summarized here was obtained with support of Ministry of Education, Youth and Sport of the Czech Republic, project nos. MSM0021620806 and 1M0021620803 and with support of the Grant Agency of the Czech Republic, project no. 304/04/0171.

References

1. Laine, R. A. (1997) The information-storing potential of the sugar code. in *Glycosciences-Status and Perspectives* (Gabius, H.-J., Gabius, S. eds.). Chapman & Hall, London, pp. 1–14.
2. Villalobo, A., Gabius, H.-J.(1998) Signaling pathways for transduction of the initial message of the glycocode into cellular responses. *Acta Anat. (Basel)* **161,** 110–129.

3. Gabius, H.-J. (2000) Biological information transfer beyond the genetic code: the sugar code. *Naturwissenschaften* **87,** 108–121.

4. Gabius, H.-J., André, S., Kaltner, H., Siebert, H.-C. (2002) The sugar code: functional lectinomics. *Biochim. Biophys. Acta.* **1572,** 165–177.

5. Gabius, H.-J., Siebert, H.-C., André, S., Jiménez, Barbero, J., Rüdiger, S. (2004) Chemical biology of the sugar code. *ChemBioChem* **5,** 740–764.

6. Stryer, L. (1990) *Biochemie.* Spektrum der Wissenschaft, Heidelberg.

7. Hounsell, E. F. (1997) Methods of glycoconjugates analysis. in *Glycosciences-Status and Perspectives* (Gabius, H.-J., Gabius, S. eds.). Chapman & Hall, London, pp. 15–29.

8. Sharon, N., Lis, H. (2004) History of lectins: from hemagglutinins to biological recognition molecules. *Glycobiology* **14,** 53R–62R.

9. Rüdiger, H., Gabius, H.-J. (2001) Plant lectins: occurrence, biochemistry, functions and applications. *Glycoconj. J.* **18,** 589–613.

10. Gabius, H.-J. (1997) Animal lectins. *Eur. J. Biochem.* **243,** 543–576.

11. Kaltner, H., Gabius, H.-J. (2001) Animal lectins: from initial description to elaborated structural and functional classification. *Adv. Exp. Med. Biol.* **491,** 79–94.

12. Kaltner, H, Stierstorfer, B. (1998) Animal lectins as cell adhesion molecules. *Acta Anat. (Basel)* **161,** 162–179.

13. Perillo, N. L., Marcus, M. E., Baum, L. G. (1998) Galectins: versatile modulators of cell adhesion, cell proliferation, and cell death. *J. Mol. Med.* **76,** 402–412.

14. Gabius, H.-J. (1997) Concepts of tumor lectinology. *Cancer Invest.* **15,** 454–464.

15. Lahm, H., André, S, Hoeflich, A., Kaltner, H., Siebert, H.-C., Sordat, B., von der Lieth, C.-W., Wolf, E., Gabius, H.-J. (2004) Tumor galectinology: insights into the complex network of a family of endogenous lectins. *Glycoconj. J.* **20,** 227–238.

16. Plzák, J., Smetana, K., Jr., Chovanec, M., Betka, J. (2005) Glycobiology of head and neck squamous epithelia and carcinomas. *Otorhinolaryngology,* **67,** 61–69.

17. Liu, F.-T., Rabinovich, G. A. (2005) Galectins as modulators of tumor progression. *Nat. Rev. Cancer* **5,** 29–41.

18. Drickamer, K. (1988) Two distinct classes of carbohydrate-recognition domains in animal lectins. *J. Biol. Chem.* **263,** 9557–9560.

19. Hirabayashi, J., Kasai, K.-I. (1991) Effect of amino acid substitution by site-directed mutagenesis on the carbohydrate recognition and stability of human 14-kDa β-galactoside-binding lectin. *J. Biol. Chem.* **266,** 23648–23653.

20. López-Lucendo, M. F., Solís, D., André, S., Hirabayshi, J., Kasai, K.-I., Kaltner, H., Gabius, H.-J., Romero, A. (2004) Growth-regulatory humangalectin-1: crystallographic characterization of the structural changes induced by single-site mutations and their impact on the thermodynamics of ligand binding. *J. Mol. Biol.* **343,** 957–970.

21. Liu, F.-T., Hsu, D. K., Zuberi, R. I., Hill, P. N., Shenhav, A., Kuwabara, I., Chen, S. S. (1996) Modulation of functional properties of galectin-3 by monoclonal antibodies binding to the non-lectin domains. *Biochemistry* **35,** 6073–6079.

22. Hirabayashi, J., Hashidate, T., Arata, Y., Nishi, N., Nakamura, T., Hirashima, M., Urashima, T., Oka, T., Futai, M., Muller, W. E., Yagi, F., Kasai, K. (2002)

Oligosaccharide specificity of galectins: a search by frontal affinity chromatography. *Biochim. Biophys. Acta.* **1572**, 232–254.

23. Liu, F.-T. (2004) Double identity: galectins may not function as lectins inside the cell. *Trends Glycosci. Glycotechnol.* **16**, 255–264.

24. Park, J. W., Voss, P. G., Grabski, S., Wang, J. L., Patterson, R. J. (2001) Association of galectin-1 and galectin-3 with Gemin4 in complexes containing the SMN protein. *Nucleic Acids Res.* **29**, 3595–3602.

25. Shimura, T., Takenaka, Y., Tsutsumi, Y., Hogan, V., Kikuchi, A., Raz, A. (2004) Galectin-3, a novel binding partner of β-catenin. *Cancer Res.* **64**, 6363–6367.

26. Chovanec, M., Smetana, K., Jr., Purkrábková, T., Holíková, Z., Dvořánková, B., André, S., Pytlík, R., Hozák, P., Plzák, J., Šedo, A., Vacík, J., Gabius, H.-J. (2004) Detection of cell-type and marker specificity of nuclear binding sites for anionic carbohydrate ligands (heparin/heparan sulfate, chondroitin sulfate, fucoidan and hyaluronic acid). *Biotech. Histochem.* **79**, 139–150.

27. Dvořánková, B., Motlík, J. Holíková, Z., Vacík, J., Smetana, K., Jr. (2002) *Dolichos biflorus* agglutinin-binding site expression in basal keratinocytes is associated with cell differentiation. *Biol. Cell.* **94**, 365–373.

28. Köttgen, E., Kage, A., Hell, B., Müller, C., Tauber, R. (1993) Glycoprotein-lectin-immunosorbent assay (GLIA). in *Lectins and Glycobiology* (Gabius, H.-J., Gabius, S. eds.). Springer Laboratory, Berlin, pp. 141–149.

29. Purkrábková, T., Smetana, K., Jr., Dvořánková, B., Holíková, Z., Böck, C., Lensch, M., André, S., Pytlík, R., Liu, F.-T., Klíma, J., Smetana, K., Motlík, J., Gabius, H.-J. (2003) New aspects of galectin functionality in nuclei of cultured bone marrow stromal and epidermal cells: biotinylated galectins as tool to detect specific binding sites. *Biol. Cell* **95**, 535–545.

30. Gabius, H.-J., Gabius, S., Zemlyanukhina, T. V., Bovin, N. V, Brinck, U., Danguy, A., Joshi, S. S., Kayser, K., Schottelius, J., Sinowatz, F., Vidal-Vanaclocha, F., Zanetta, J. P. (1993) Reverse lectin histochemistry: design and application of glycoligands for detection of cell and tissue lectins. *Histol. Histopathol.* **8**, 369–383.

31. Bovin, N. V. (2002) Neoglycoconjugates: trade and art. *Biochem. Soc. Symp.* **69**, 143–160.

32. André, S., Ortega, P. J., Perez, M. A., Roy, R., Gabius, H.-J. (1999) Lactose-containing starburst dendrimers: influence of dendrimer generation and binding-site orientation of receptors (plant/animal lectins and immunoglobulins) on binding properties. *Glycobiology* **9**, 1253–1261.

33. Bovin, N. V. (1993) Sugar polyacrylamide conjugates as probes for cell lectins. in *Lectins and Glycobiology* (Gabius, H.-J., Gabius, S. eds.). Springer Laboratory, Berlin, pp. 23–28.

34. Froňková, V., Holíková, Z., Liu, F.-T., Homolka, J., Rijken, D. C., André, S., Bovin, N. V., Smetana, K., Jr., Gabius, H.-J. (1999) Simultaneous detection of endogenous lectins and their binding capacity at the single-cell level-a technical note. *Folia Biol. (Praha)* **45**, 157–162.

35. Plzák, J., Holíková, Z., Dvoránková, B., Smetana, K., Jr., Betka, J., Hercogová, J., Saeland, S., Bovin, N. V., Gabius, H.-J. (2002) Analysis of binding of mannosides

in relation to Langerin (CD207) in Langerhans cells of normal and transformed epithelia. *Histochem. J.* **34**, 247–253.

36. John, C. M., Leffler, H., Kahl-Knutsson B., Svensson, I., Jarvis, G. A., John, C. M., Leffler, H., Kahl-Knutsson, B., Svensson, I., Jarvis, G. A. (2003) Truncated galectin-3 inhibits tumor growth and metastasis in orthotopic nude mouse model of human breast cancer. *Clin. Cancer. Res.* **9**, 2374–2383.

37. Smetana, K., Jr., Plzák, J., Dvoránková, B., Holíková, Z. (2003) Functional consequences of glycophenotype of squamous epithelia – practical employment. *Folia Biol. (Praha)* **49**, 118–127.

38. Chovanec, M., Smetana, K., Jr., Dvořánková, B., Plzáková, Z., André, S., Gabius, H.-J. (2004) Decrease of nuclear reactivity to growth-regulatory galectin-1 in senescent human keratinocytes and detection of non-uniform staining profile alterations upon prolonged culture for galectins-1 and -3. *Anat. Histol. Embryol.* **33**, 348–354.

39. Klíma, J., Smetana, K., Jr., Motlík, J., Plzáková, Z., Liu, F.-T., Štork, J., Kaltner, H., Chovanec, M., Dvořánková, B., André, S., Gabius, H.-J. (2005) Comparative phenotypic characterization of keratinocytes originating from hair follicles. *J. Mol. Histol.* **36**, 89–96.

40. Plzák, J., Smetana, K., Jr., Betka, J., Kodet, R., Kaltner, H., Gabius, H.-J. (2000) Endogenous lectins (galectins-1 and -3) as probes to detect differentiation-dependent alterations in human squamous cell carcinomas of oropharynx and larynx. *Int. J. Mol. Med.* **5**, 369–372.

41. Konstantinov, K. N., Shames, B., Izuno, G., Liu, F.-T. (1994) Expression of epsilon BP, a beta-galactoside-binding soluble lectin, in normal and neoplastic epidermis. *Exp. Dermatol.* **3**, 9–16.

42. Plzák, J., Smetana, K., Jr., Hrdlicková, E., Kodet, R., Holíková, Z., Liu, F.-T., Dvořánková, B., Kaltner, H., Betka, J., Gabius, H.-J. (2001) Expression of galectin-3-reactive ligands in squamous cancer and normal epithelial cells as a marker of differentiation. *Int. J. Oncol.* **19**, 59–64.

43. Betka, J., Plzák, J., Smetana, K., Jr., Gabius, H.-J. (2003) Galectin-3, an endogenous lectin, as tool to monitor cell differentiation in head and neck carcinomas with implications for lectin-glycan functionality. *Acta Otolaryngol.* **123**, 261–263.

44. Holíková, Z., Hrdličková-Cela, E., Plzák, J., Smetana, K., Jr., Betka, J., Dvořánková, B., Esner, M., Wasano, K., André, S., Kaltner, H., Motlík, J., Hercogová, J., Kodet, R., Gabius, H.-J. (2002) Defining the glycophenotype of squamous epithelia by plant and mammalian lectins. Differentiation-dependent expression of α2,6- and α2,3-linked N-acetylneuraminic acid in squamous epithelia and carcinomas and its differential effect on binding of the endogenous lectins galectins-1 and -3. *APMIS* **110**, 845–856.

45. Choufani, G., Nagy, N., Saussez, S., Marchant, H., Bisschop, P., Burchert, M., Danguy, A., Louryan, S., Salmon, S., Gabius, H.-J., Kiss, R., Hassid, H. (1999) The levels of expression of galectin-1, galectin-3, and the level of Thomsen-Friedenreich antigen and their binding sites decrease as clinical aggressiveness increases in head and neck cancers. *Cancer* **86**, 2353–2363.

46. Plzák, J., Betka, J., Smetana, K., Jr., Chovanec, M., Kaltner, H., André, S., Kodet, R., Gabius, H.-J. (2004) Lectin histochemistry with an endogenous adhesion/growth-modulatory protein galectin-3 as emerging technique in prognostic evaluation: case study for advanced head and neck carcinoma. *Eur. J. Cancer.* **40**, 2324–2330.

47. Magnaldo, T., Fowlid, D., Darmon, M. (1998) Galectin-7, a marker of all types of stratified epithelia. *Differentiation* **63**, 159–168.

48. Bernerd, F., Sarasin, A., Magnaldo, T. (1999) Galectin-7 overexpression is associated with the apoptotic processes in UVB-induced sunburn keratinocytes. *Proc. Natl. Acad. Sci. U. S. A.* **96**, 11329–11334.

49. Chovanec, M., Smetana, K., Jr., Plzák, J., Betka, J., Plzáková, Z., Štork, J., Hrdličková, E., Kuwabara, I., Dvořánková, B., Liu, F.-T., Gabius, H.-J. (2005) Detection of new diagnostic markers in pathology by focus on growth-regulatory endogenous lectins. The case study of galectin-7 in squamous epithelia, *Prague Med. Rep.* **106**, 209–216.

50. Ueda, S., Kuwabara, I., Liu, F.-T. (2004) Suppression of tumor growth by galectin-7 gene transfer. *Cancer Res.* **64**, 5672–5676.

51. Plzák, J., Holíková, Z., Smetana, K., Jr., Riedel, F., Betka, J. (2003) The role of dendritic cells in the pharynx. *Eur. Arch. Otorhinolaryngol.* **260**, 266–272.

52. Smetana, K., Jr., Holíková, Z., Klubal, R., Bovin, N. V., Dvořánková, B., Bartůnková, J., Liu, F.-T., Gabius, H.-J. (1999) Coexpression of binding sites for A(B) histo-blood group trisaccharides with galectin-3 and Lag antigen in human Langerhans cells. *J. Leukoc. Biol.* **66**, 644–649.

53. Holíková, Z., Smetana, K., Jr., Bartůnková, J., Dvořánková, B., Kaltner, H., Gabius, H.-J. (2000) Human epidermal Langerhans cells are selectively recognized by galectin-3 but not by galectin-1. *Folia Biol.(Praha)* **46**, 195–198.

54. Valladeau, J., Ravel, O., Dezutter-Dambuyant, C., Moore, K., Kleijmeer M., Liu, Y., Duvert-Frances, V, Vincent, C., Schmitt, D., Davoust, J., Caux, C., Lebecque, S., Saeland, S. (2000) Langerin, a novel C-type lectin specific to Langerhans cells, is an endocytic receptor that induces the formation of Birbeck granules. *Immunity* **12**, 71–81.

16

A Multiplex Approach to Isotyping Antigen-Specific Antibodies Using Biotinylated Antigen/Streptavidin-Phycoerythrin

Krista McCutcheon

Summary

Analytical methods characterizing the immunogenicity of antigens are useful for monitoring, characterizing and predicting antibody responses to therapeutic biologics or vaccines. Distinct Luminex® microspheres coupled with protein G, anti-human immunoglobulin (Ig)A, anti-human IgM and anti-human IgE were developed for the simultaneous capture of total IgG, IgA, IgM and IgE (IgGAME) antibodies from human or non-human primate serum. The fraction of antigen-specific antibodies captured on the beads was detected using biotinylated antigen/streptavidin-phycoerythrin. The method was demonstrated by isotyping antibodies directed against an anti-CD11a antibody therapeutic (RAPTIVA®/efalizumab) from the serum of a cynomolgus monkey hyper-immunized with RAPTIVA® over a 15-month period. The quantitative range of the antibody measurements, using 5 µl of sample, was determined to be 15 ng/ml to 50 µg/ml in 10% serum. By the use of any biotinylated antigen as a detector, this multiplexed isotyping assay can be broadly applied to human and non-human primate IgGAME immunogenicity studies.

Key Words: Immunogenicity; IgG; IgA; IgM; IgE; isotype; antibody; antigen; multiplexing.

1. Introduction

The immunogenicity of protein therapeutics leading to the production of specific antibodies has been correlated with altered therapeutic pharmacokinetics, potency, and adverse events (*1*). Assays screening for antibodies against therapeutic biologicals is a requirement during most clinical trials and may be permanently implemented for long-term patient monitoring in cases where a

From: *Methods in Molecular Biology, vol. 418: Avidin-Biotin Interactions, Methods and Applications*
Edited by: R. J. McMahon © Humana Press, Totowa, NJ

significant risk has been demonstrated. When a subject develops antibodies to a therapeutic, follow-up characterization of the antibodies can include measurements of titer, specificity and cross-reactivity, affinity and isotype *(2)*.

In studies of vaccine efficacy and infectious diseases, the beneficial protective effects of particular antibody isotypes have been reported *(3)*. However, the clinical significance of antibody isotypes in unwanted immunogenicity to therapeutic products is not well understood. Each of the immunoglobulin (Ig)G, IgA, IgM and IgE antibody classes has its own time course of appearance, circulating serum concentration, half-life, structure, and interactions with other molecules and cells of the immune system, such as complement and effector cells. IgA pre-dominants in mucosal immunity, IgE is associated with allergic reactions, and the presence of IgM suggests a less mature immune response. Factors that could each affect the magnitude and class of an antibody response include the route of therapeutic administration, whether it be oral, inhaled, intravenous or intramuscular injection; the site of therapeutic accumulation; the target tissue; the mechanism and site of therapeutic clearance in vivo; and the patient population.

Because the volume of patient serum available for characterization assays can be limiting, it is advantageous to have a multiplexed assay able to measure multiple parameters in one small aliquot. Luminex® technology uses uniquely color-coded 5-µm microspheres to identify multiple reactions in a single tube or well. Using this method over 100 distinct microspheres can be created by coupling each capture antibody of interest to a bead population by amine chemistry. Sample analyte is captured and a detector with a fluorescent label [phycoerythrin (PE)] creates a sandwich complex used to quantify by intensity the amount of analyte captured. In this report, antigen-specific Igs were isotyped in a 4-plex [IgG, IgA, IgM and IgE (IgGAME)] bead array using a combined biotinylated antigen/streptavidin–PE detection system. The Luminex® instrument operates similar to traditional FACS methods. Hydrodynamic focusing aligns the beads in single file and two lasers interrogate them, one at a time. The first laser excites the fluorescent detector to quantify the analyte. The second laser excites the bead to identify its number. Because all the beads are the same size, doublet discrimination can be used to eliminate bead aggregates, ensuring only single bead measurements are reported. Typically, assays are set up with cocktails containing separate standards or controls for each analyte in the assay. This is because while run in multiplex, the capture and detection properties for each analyte retain individual sensitivity, recovery and standard curve characteristics.

A source of positive controls for antigen-specific IgGAME was needed to develop the multiplexed isotyping assay. Cynomolgus monkeys are physiologically similar to humans, and thus are commonly used as an animal model for therapeutic safety and efficacy studies. Cynomolgus monkeys are also frequently hyper-immunized with biological therapeutics to generate polyclonal antibody positive controls for use in human anti-therapeutic antibody screening assays developed to support safety and efficacy monitoring during clinical studies. Thereby, hyper-immune serum containing anti-RAPTIVA® IgGAME antibodies was identified and used as a source of positive controls for this study. RAPTIVA® (efalizumab), an anti-CD11a recombinant human antibody, was FDA approved in 2003 for chronic moderate-to-severe plaque psoriasis in adults.

2. Materials

2.1. Antibodies

1. Mouse anti-human IgA antibody, clone IgA5-3B, was purchased from Custom Monoclonals International (Sacramento, CA), and rat anti-human IgE, clone LO-HE-17, was purchased from BioSource (Camarillo, CA). These antibodies were coupled in-house to microspheres for use as capture reagents (*see* **Note 1**).
2. Microspheres coupled with anti-human IgM (product number 42–315) and protein G (product number 42–006A) were purchased from the Beadylyte™ product line by Upstate Inc., (Charlottesville, PA) (*see* **Note 2**).
3. Purified IgGAME human myeloma Igs for testing assay specificity were purchased from Calbiochem (San Diego, CA).
4. Antibodies biotinylated for use as detectors included a cocktail of Beadlyte™ anti-human kappa light chain and anti-human lambda chain antibodies (product 44–300, Upstate Inc.) and RAPTIVA® (Genentech, Inc., South San Francisco, CA) (*see* **Note 3**).

2.2. Anti-Therapeutic Hyper-Immune Serum

1. A cynomolgus monkey was administered every 2 weeks with 1 mg in 0.5 ml of RAPTIVA®, plus adjuvant, subcutaneously in the neck. For the primary immunization, TiterMax was used, and Freund's Incomplete Adjuvant was used for the subsequent boosts. The immunization period was 15 months. RAPTIVA® is known to show no binding to cynomolgous monkey CD11a.

2.3. Coupling of Antibody to Microspheres

1. Anti-human IgA or IgE antibodies or RAPTIVA®, diluted in coupling buffer (*see* **Note 4**).
2. Luminex Carboxylated Microspheres (Luminex Corp., Austin, TX) (*see* **Note 5**).
3. Eppendorf tubes, 1.5 mL (USA Scientific, Inc., Ocala, FL) (*see* **Note 6**).
4. Multiscreen-BV, 96-well filter plates (MABVN1250, Millipore, Bedford, MA) (*see* **Note 7**).

5. Millipore Multiscreen Vacuum Manifold Apparatus (Millipore).
6. Activation buffer: 0.1 M sodium phosphate (pH 6.1 $\pm$ 0.1). Prepare 1 M stocks of dibasic (Na$_2$HPO$_4$; 142 g/l) and monobasic (NaH$_2$PO$_4$; 120g/l) phosphate buffers in deionized water without any pH adjustment. Combine 12 ml of 1 M Na$_2$HPO$_4$ with 88 ml of 1 M NaH$_2$PO$_4$ in a final volume of 1 l deionized water (no pH adjustment).
7. Coupling buffer: Phosphate-buffered saline (PBS) (pH 7.3 $\pm$ 0.1). Dissolve 8 g NaCl, 0.2 g KCl, 1.44 g Na$_2$HPO$_4$ and 0.24 g KH$_2$PO$_4$ in 800 ml deionized water. Adjust pH with HCl and add 1 l water.
8. Wash buffer: PBS (pH 7.3 $\pm$ 0.1), 0.05% polysorbate-20. Add 0.5 ml polysorbate-20 to 1 l of PBS (*see* **Note 8**).
9. Blocking/storage buffer: PBS (pH 7.3 $\pm$ 0.1), 1 mg/ml bovine serum albumin (BSA) (Fraction V).
10. Assay diluent: PBS, 0.5% BSA, 0.05% polysorbate-20. BSA (Fraction V) (5 g) and 0.5 ml of polysorbate-20 are added to 1 l PBS.
11. Sulpho-NHS: N-Hydroxysulfosuccinimide sodium salt (Pierce Biotechnology, Inc., Rockford, IL).
12. EDC: 1-Ethyl-3-(3-demethyaminopropyl)carbodiimide-HCl) (Pierce Biotechnology, Inc.).
13. Microbead sonicator water bath (Upstate Inc.).
14. Centrifugation: Microcentrifuge, all steps at setting 11.5 (about 10,000 *g*).
15. Hemacytometer (VWR Int., Bristol, CT).

2.4. Biotinylation of Antigen

1. RAPTIVA® in PBS at a concentration of 1 mg/ml (*see* **Note 4** on removal of interfering substances).
2. NHS-Biotin, Pierce EZ-link (product 20217, Pierce Biotechnology, Inc.).
3. Sodium bicarbonate buffer: NaHCO$_3$, pH 8.0–8.6. Dissolve 8.4 g NaHCO$_3$ powder in 1 l of deionized water, do not adjust pH).
4. PBS, pH 7.3 $\pm$ 0.1. Dissolve 8 g NaCl, 0.2 g KCl, 1.44 g Na$_2$HPO$_4$ and 0.24 g KH$_2$PO$_4$ in 800 ml deionized water. Adjust pH with HCl and add water to 1 l.
5. Dimethyl sulfoxide (DMSO, Sigma, St. Louis, MO).
6. Dialysis membrane, 10,000 molecular weight cutoff (e.g. 10,000 MWCO Slide-A-Lyzer®, Pierce Biotechnology, Inc.) (*see* **Note 9**).

2.5. Preparation of Affinity Sepharose and Purification of Antigen-Specific Serum Antibodies

1. Activated CH-sepharose 4B (Amersham Biosciences, Uppsala) (*see* **Note 10**).
2. Chromatography column (e.g., disposable polystyrene columns, Pierce product 29920).
3. Ice-cold HCl, 1 mM: 11.8 µL 10.6 M HCl in 125 ml deionized water.
4. Coupling buffer: 0.1 M NaHCO$_3$, pH 8, 0.5 M NaCl. Dissolve 8.4 g NaHCO$_3$ powder and 29.22 g NaCl in 1 l of deionized water, do not adjust pH).

5. RAPTIVA®, 1 mg/ml in coupling buffer (*see* **Note 4** on removal of interfering substances).
6. Tris buffer, 0.1 M, pH 8: Tris (hydroxymethylaminomethane), 12.11 g, is added to 1 l of deionized water and adjusted with HCl to pH 8.
7. Low pH wash buffer: acetic acid added dropwise to 0.5 M NaCl in deionized water until pH 4 is reached.
8. PBS, pH 7.3 ± 0.1. Dissolve 8 g NaCl, 0.2 g KCl, 1.44 g Na_2HPO_4 and 0.24 g KH_2PO_4 in 800 ml deionized water. Adjust pH with HCl and add water to 1 l.
9. Elution buffer: 0.1 M glycine, pH 2.3. Glycine, 3.75 g, is added to 450 ml of deionized water, adjusted to pH 2.3 with HCl, and the volume made up to 500 ml.
10. Tris buffer ,1 M, pH 8: Tris (hydroxymethylaminomethane), 12.11 g, is added to 100 ml of deionized water and adjusted with HCl to pH 8.

2.6. Multiplexed IgGAME Assay

1. Assay diluent: PBS, 0.5% BSA, 0.05% polysorbate-20. BSA (Fraction V) (5 g) and 0.5 ml of polysorbate-20 are added to 1 l of PBS.
2. IgGAME bead working solution: Bead stock vials are sonicated for 1–2 min and vortexed. 5×10^5 beads/ml of each of the four bead populations are mixed in assay diluent, vortexed and sonicated for 1–2 min just before adding to plate (*see* **Note 11**).
3. Biotinylated RAPTIVA®, prepared at 5 µg/ml in assay diluent (*see* **Note 12**).
4. Streptavidin-PE (Molecular Probes product S-866, Eugene, OR), prepared at 2.5 µg/ml in assay diluent.
5. Wash buffer: PBS, containing 0.05% polysorbate-20 (0.5 ml/l of PBS).
6. Fix buffer: 5% formaldehyde in PBS: Dilute 1.35 ml of 37% formalin in a final volume of 10 ml PBS.
7. Ninety-six-well RIA plates, flat bottom (Costar, Corning product 3591, VWR, Bristol, CT).
8. Multiscreen-BV, 96-well filter plates (MABVN1250, Millipore).
9. Millipore Multiscreen Vacuum manifold apparatus (Millipore).
10. Microbead sonicator water bath (Upstate Inc.)
11. Luminex® instrument, sheath buffer and CAL1 and CAL2 calibrators (*see* **Note 13**).

3. Methods

A multiplex assay was developed for simultaneously detecting anti-RAPTIVA® IgGAME antibodies from a 5 µl serum sample at a minimum dilution of 1/10. Four distinct Luminex® microspheres were coupled with protein G, anti-human IgA, anti-human IgM or anti-human IgE. A protocol was optimized using unpurified anti-RAPTIVA® cynomolgus monkey serum and biotinylated RAPTIVA®/streptavidin-PE for detection. Singleplex measurements of total antibody responses, irrespective of isotype, were performed by capture of serum antibodies on beads coupled with RAPTIVA®, followed by the same biotinylated RAPTIVA®/streptavidin-PE method of detection (a bridging format).

In most assays, biotinylated RAPTIVA®/streptavidin-PE was used for detection. However, for characterization of each capture bead's isotype cross-reactivity and sensitivity properties, a singleplex format detected by a cocktail of biotinylated anti-kappa/lambda chain antibodies and streptavidin-PE was used. Capture beads were first tested for specificity using purified myeloma human Igs in assay diluent. Data are shown in relative fluorescent units (RFU) with buffer background subtracted (see Table 1). Standard curves for each of the four beads were tested on an IgGAME cocktail of purified human Igs to demonstrate a dose-dependent specificity of the signal (not shown). Second, capture beads were tested for cross-reaction to monkey Igs using total serum Ig measurements before and after immunization using biotinylated RAPTIVA®/streptavidin-PE detection. Data are shown as RFU with buffer background subtracted (see Fig. 1). Because monkey Igs are purified from serum they are typically contaminated with other isotypes and therefore not useful for specificity studies. Third, the signal from the beads was tested and found to be comparable in singleplex and four-plex formats, using biotinylated RAPTIVA®/streptavidin-PE detection (see Table 2).

The coupling of capture antibody to the bead is an important step for optimizing sensitivity. The optimal antibody coupling concentration was decided based on the amount giving a maximum signal from RAPTIVA®, immune sera (see Fig. 6). Although for all antibody coupling the values would be quite similar to each other (2.5–5 µg protein/million beads, repre-

Table 1
Specificity of Capture Beads Measured with Purified Human Immunoglobulins (Igs)

Immunoglobulin	Median reactive fluorescent unit for capture bead[a]			
	Protein G	Anti-IgA	Anti-IgM	Anti-IgE
Human IgG	15811	0	23	11
Human IgA	0	3905	34	0
Human IgM	0	29	4879	7
Human IgE	0	0	0	12118

[a]Detected with a cocktail of biotinylated anti-human kappa/lambda light chain antibodies/ streptavidin-phycoerythin.

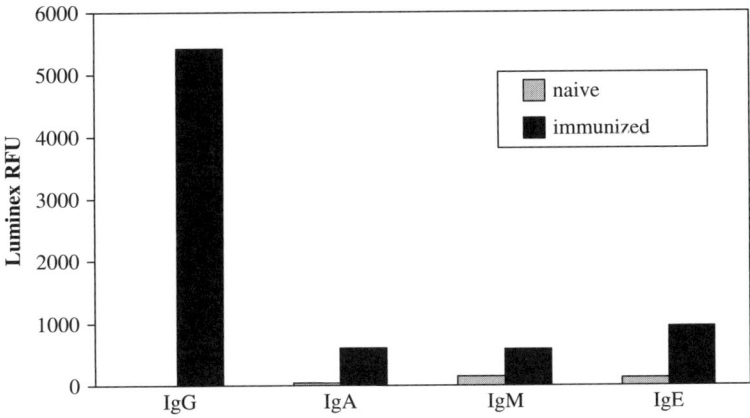

Fig. 1. Development of capture reagents cross-reactive to cynomolgus monkey immunoglobulins(Igs). Day 210 RAPTIVA®-immune serum was tested against naïve serum, using biotinylated RAPTIVA®/streptavidin-phycoerythin detection. The capture beads showing the best median relative fluorescent unit (RFU) signal-to-noise ratio were selected.

senting approximately 12 IgG protein molecules per 100 COOH groups in the coupling reaction) each capture bead should be optimized in this range. The number of beads input per analyte was selected at 5000 to minimize bead carry-over well to well, while maximizing sensitivity and speed of acquisition (*see* **Fig. 7**).

The sensitivity of the protein G and RAPTIVA® capture beads could be assessed using affinity purified anti-RAPTIVA® serum, spiked at known and increasing concentrations into 10% naïve cynomolgus monkey serum,

Table 2
Singleplex Versus Multiplex Assay Performance Measured using RAPTIVA®, Day 210 Hyper-Immune or Naive (for background) Cynomolgus Monkey Serum

Microsphere	Singleplex RFU[a] (background)	Multiplex RFU[a] (background)	% difference (after background subtracted)
Protein G	19541 (3894)	20316 (4387)	2
Anti-IgA	885 (40)	881 (46)	1
Anti-IgM	126 (60)	131 (59)	9
Anti-IgE	3206 (93)	3224 (109)	0

[a]RFU, relative fluorescent unit. Detected with biotinylated RAPTIVA®/Streptavidin-phycoerythin.

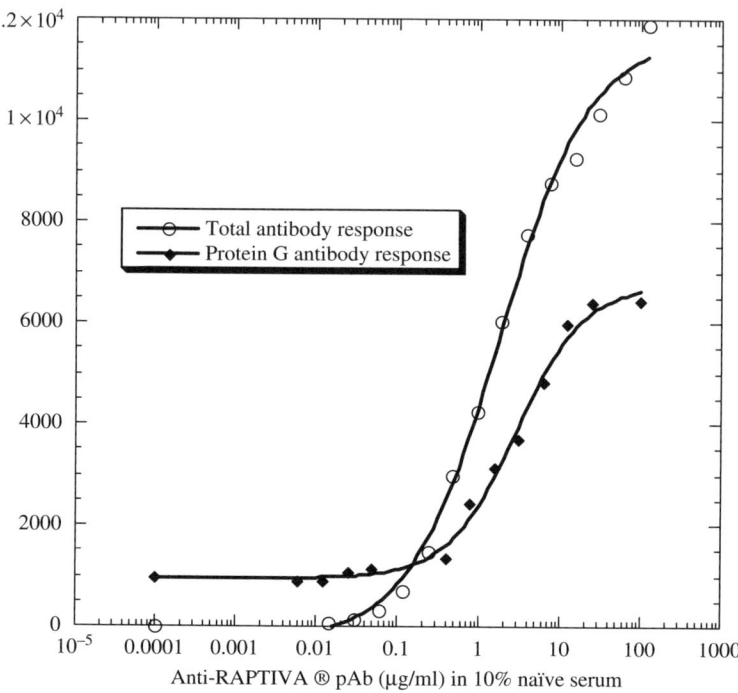

Fig. 2. Sensitivity of the total and immunoglobulin(Ig)G antibody assays. Affinity purified pAb to RAPTIVA® was spiked at known concentrations into naive, 10% cynomolgus monkey serum. RAPTIVA® or protein G capture beads were added and the assay performed as described in the methods using biotinylated RAPTIVA®/streptavidin-phycoerythin detection. Standard curves were also generated in single-plex for each of the four capture beads using a cocktail of purified human Ig standards (not shown) and biotinylated anti-Ig light chain detectors.

followed by detection with biotinylated RAPTIVA® (*see* **Fig. 2**). The sensitivity of the total antibody assay was in the range of 50 ng/ml, whereas specific IgG antibodies, in 10% serum, could be detected in the range of 0.2 µg/ml, representing about 0.02% of total serum IgG. The lower sensitivity of the IgG capture is due to the high concentration of non-specific serum IgG [average 15 mg/ml (*4*)] competing for the capture of drug-specific IgG on protein G.

Measuring the anti-RAPTIVA® IgGAME response in serum over a 15-month period showed the assay to be sensitive to changes in the isotype of the immune response over time (*see* **Fig. 3**). The cynomolgus monkeys used for the development of this assay were immunized frequently in the presence of

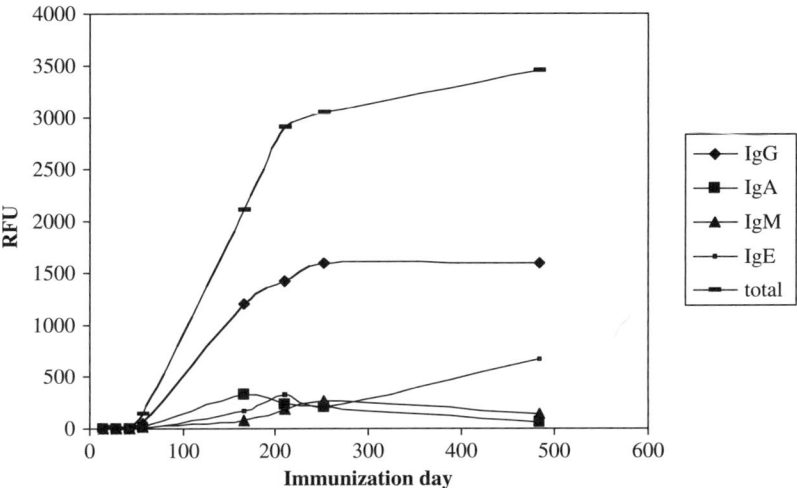

Fig. 3. Time course of the anti-RAPTIVA® immunoglobulin(Ig)G, IgA, IgM and IgE IgGAME response. Serum from day 14 to day 489 of a cynomolgus monkey hyper-immunized with therapeutic was measured at a 1/10 dilution in the IgGAME and total antibody assays using biotinylated RAPTIVA® /streptavidin-phycoerythrin detection.

adjuvant with the therapeutic in order to elicit an immune response and thus results are not indicative of what would be expected in pre-clinical study dosing regimes.

The specificity of the total and IgE anti-RAPTIVA® immune responses was tested by competing with a panel of Genentech IgG therapeutics with similar frameworks, and generic human IgG and FAb´2 molecules (*see* **Figs 4** and **5**). The results indicate that 50% of the total antibody immune response was RAPTIVA® specific. As other Genentech IgG therapeutics are nearly identical in sequence outside of the complementary-determining region, the total RAPTIVA®-specific immune response was thought to be predominantly directed to the antigen-binding regions. Interestingly, the IgE fraction of anti-RAPTIVA® antibodies appeared to be directed entirely to framework sequences, particularly in the hinge region (as the assay signal was equally competed by any human IgG or F(ab´)2 portion). Sequence alignments of published IgG molecules revealed the hinge region to be the most divergent area of Ig sequence between cynomolgous monkeys and humans.

Finally, the practicability of the method was demonstrated by evaluating the intra-assay and inter-assay precision and the stability of measurements in freeze/thawed serum samples stored over time (*see* **Tables 3** and **4**).

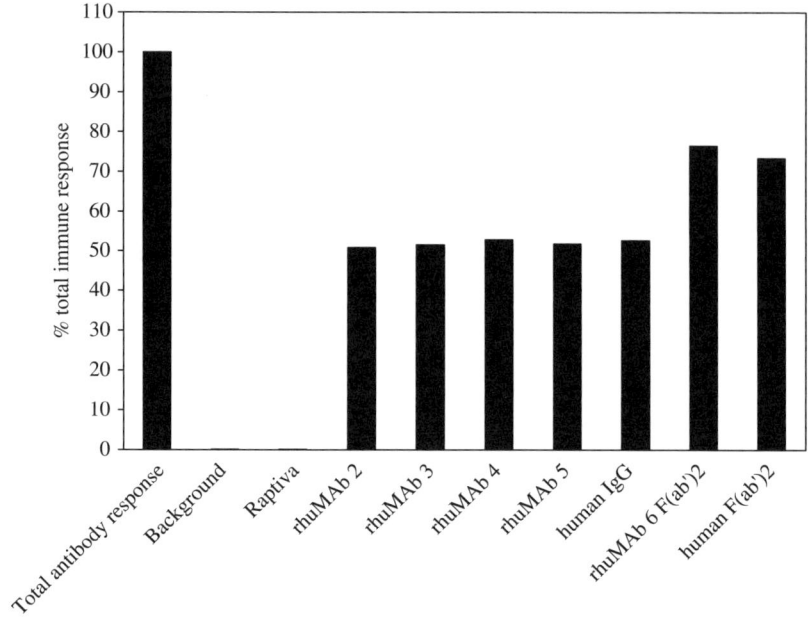

Fig. 4. Specificity of the total anti-RAPTIVA® antibody response. A single-plex, bridging format was used to measure the total antibody response, irrespective of isotype, wherein RAPTIVA® was used both on the capture bead and as a detector with streptavidin-phycoerythin. Therapeutic-specific immunoglobulin(Ig) was reported as a percentage of the Day 210 median reactive fluorescent unit. Anti-RAPTIVA® hyper-immune serum was pre-incubated for 2 h at room temperature, at a dilution of 1/10, with 10 µg/ml of the immunizing humanized monoclonal antibody (RAPTIVA®), other humanized monoclonal antibodies of differing complementary-determining region (rhuMAb2-5), commercially available human IgG or F(ab′)2.

3.1. Conjugation of Capture Microspheres

3.1.1. Preparation

1. Allow all reagents to warm to room temperature.
2. Dilute the Ig stock in coupling buffer to 50 µg/ml in 0.125 ml (small scale) or 31.25 µg/ml in 1ml (large scale) (*see* **Note 14**).
3. Weigh approximately 5 mg of Sulfo-NHS into an Eppendorf tube. Repeat for EDC. Make these immediately before use and minimize exposure to air. Store dry reagents in a desiccator at –20°C.

3.1.2. Microsphere Activation

1. Centrifuge the Luminex® microsphere stock for 10 s.
2. Disperse the microsphere pellet with 2 min of sonication, and vortex for 20 s.

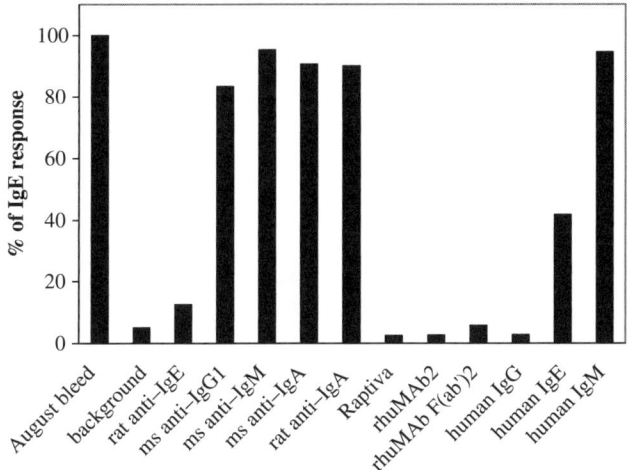

Fig. 5. Specificity of the immunoglobulin(Ig)E anti-RAPTIVA® antibody response. Total cynomolgus monkey IgE antibodies were captured from August, (Day 210), hyper-immune serum, with anti-IgE beads. Therapeutic-specific IgE was reported as a percentage of the Day 210 median reactive fluorescent unit, using biotinylated RAPTIVA® followed by streptavidin- phycoerythrin. The IgE capture specificity was confirmed using 100 µg/ml of anti-IgG, anti-IgA, anti-IgM and anti-IgE monoclonal antibodies, incubated with immune serum at a dilution of 1/10 in assay diluent, overnight at room temperature. The IgE framework specificity was shown by pre-incubation of immune serum with 10–12 µg/ml of several humanized monoclonal IgG or F(ab')2 antibodies, or commercially available human IgG, IgM or IgE, for 2 h at room temperature.

3. Dispense 2.5×10^6 microspheres (1.25×10^7 beads for large scale) from the stock into a USA Scientific Eppendorf tube (*see* **Note 6**).
4. Centrifuge for 1 min and aspirate supernatant.
5. Resuspend the microspheres in 80 µl (400 µl for large scale) of room temperature activation buffer and sonicate for 1 min or until a homogeneous mixture of microspheres is observed.
6. Immediately before use make a 50 mg/ml Sulfo-NHS solution in activation buffer (add 100 µl of activation buffer to pre-weighed 5 mg of Sulfo-NHS). Also, make a 5 mg/ml (10 mg/ ml for large scale) EDC solution in activation buffer (add 1 ml activation buffer to pre-weighed 5 mg of EDC or 0.5 ml for large scale).
7. Add 10 µl (50 µl for large scale) of the Sulpho-NHS solution to the microsphere suspension and vortex gently. Immediately add 10 µL (25 µl for large scale) of the EDC solution and vortex gently.
8. Incubate the suspension for 20 ± 2 min in the dark at room temperature (e.g., in a drawer).

Table 3
Immunoglobulin (Ig) GAME Intra-Assay and Inter-Assay Precision Over 4 Days Using Day 166, RAPTIVA®Hyper-Immune Cynomolgus Monkey Serum

Assay		IgG	IgA	IgM	IgE	Total
	Day 166 serum (1/10) IgGAME data: intra-assay and inter-assay					
1	RFU	3430	351	133	232	4140
	Intra- % CV	4.1	1.6	3.2	3.7	2.8
2	RFU	3793	420	74	319	5465
	Intra- % CV	1.6	2.2	1.0	2.6	0.4
3	RFU	3722	389	144	298	5402
	Intra- % CV	4.0	4.2	1.4	1.4	4.3
4	RFU	3837	445	86	368	4945
	Intra- % CV	2.0	7.8	1.6	1.5	1.1
Mean	RFU	3695	401	109	304	4988
	Inter-SD	183.2	40.6	34.3	56.4	611.0
	Inter-% CV	5.0	10.1	31.4	18.4	12.2

IgGAME, IgG, IgA, IgM and IgE; RFU, relative fluorescent unit; CV, coefficient of variation; SD, standard deviation.

Table 4
Stability of Immunoglobulin (Ig) GAME Measurements of Day 166, RAPTIVA®Hyper-Immune Cynomolgus Monkey Serum During −70°C Thaws and 4°C Storage.

Storage		IgG	IgA	IgM	IgE	Total
	Day 166 Serum (1/10) IgGAME data: analyte stability					
1× thaw	RFU	3837	445	86	368	4945
	intra-% CV	2.0	7.8	1.6	1.5	1.1
	% difference	0	0	0	0	0
2× thaw	RFU	3880	436	93	383	5173
	intra-% CV	1.2	2.3	0.8	4.8	2.0
	% difference	1.1	−2.0	8.1	4.1	4.6
3× thaw	RFU	3849	462	89		5173
	intra-% CV	1.1	2.6	7.2	0.5	2.5
	% difference	0.3	3.8	3.5	13.3	7.9
4°C	RFU	4196	978	177	87	11590
4 weeks	intra-% CV	3.0	3.0	8.5	3.9	0.0
	% difference	9.4	119.8	105.8	-76.4	134.4

IgGAME, IgG, IgA, IgM and IgE; RFU, relative fluorescent unit; CV, coefficient of variation.

9. Apply the activated microspheres to a filter plate and remove the buffer by vacuum. Wash the beads in the filter plate 2× 250 µl (2× 500 µl for large scale) coupling buffer. Remove the beads in 125 µl of coupling buffer into an Eppendorf tube. Immediately go to coupling steps (*see* **Note 7**).

3.1.3. Coupling, Blocking and Storage

1. To initiate coupling, add 125 µl of 50 µg/ml anti-Ig antibody or RAPTIVA®, (1 ml of 31.25 µg/ml for large scale) preparation to the 125 µl of beads. Mix by gentle vortexing.
2. Sonicate for 1 h in the dark at room temperature. Heat generated during this sonication time is acceptable.
3. Centrifuge the protein-coupled microspheres for 1 min and aspirate the supernatant carefully (filter plate washing not necessary at this step).
4. Wash the beads 2× 250 µl PBS (2× 500 µl for large scale) with centrifugation and careful aspiration.
5. To block, add 250 µl (1 ml for large scale) of blocking/storage buffer, mix by gentle vortexing, followed by a 1-min sonication.
6. Incubate at room temperature in the dark for 30 min.
7. Centrifuge the coupled and blocked microspheres for 1 min and aspirate supernatant.
8. Suspend pellet in desired volume of blocking/storage buffer and enumerate the microsphere preparation on a hemacytometer.
9. Store at 2–8°C, protected from light for up to or > 1 year. Working stocks at 5 × 10^5 beads/ml in assay diluent are prepared for use in assays (so that 10 µl of each bead type gives 5000 beads per well in an assay) (*see* **Note 15**).

3.2. Biotinylation of Monoclonal Antibodies

1. Concentrate antibody by ultrafiltration or lyophilization if concentration is <0.5 mg/ml prior to the start of this protocol. Dialyze the desired amount of purified antibody versus 0.1 M $NaHCO_3$ buffer, pH 8.0–8.6, for at least 2 h at room temperature.
2. Transfer dialyzed antibody to an Eppendorf tube and adjust concentration to 0.5–1.2 mg/ml with $NaHCO_3$ if needed. For consistency, using the same protein concentration from batch to batch is preferable.
3. Calculate total milligrams of antibody by dividing OD_{280} by the approximate IgG extinction coefficient value of 1.4, then multiplying by the volume in the tube. Milligrams of antibody = (OD_{280}) (volume in tube)/1.4. Alternatively, a colorimetric protein assay using an IgG or BSA standard curve can be used.
4. Dissolve 2 mg of NHS-Biotin in 1 ml of DMSO.
5. Add 100 µg (50 µl of stock) biotin per mg of antibody, vortex gently, and allow to react at room temperature for at least 2 h. This represents a molar challenge ratio of 27 biotin : 1 IgG, which will typically result in the incorporation of 6–12 biotin/molecule (*see* **Note 16**).

6. Transfer to dialysis cassette and dialyze against PBS (2 l) with one buffer change after 2 h. Dialyze overnight at 2–8°C.
7. Determine final concentration by protein assay or OD_{280}.
8. Aliquot and store at –20 to –70°C.

3.3. Purification of Antigen-Specific Antibodies from Serum

3.3.1. Preparation of Affinity Matrix

1. Weigh out 0.3 g of activated CH-sepharose 4B beads (gives about 1 ml bed volume).
2. Wash and swell the beads in a 10-ml column with 1 mM ice-cold HCl.
3. Equilibrate the beads with 10 ml coupling buffer.
4. Mix the ligand (1 mg RAPTIVA® in 5 ml coupling buffer) and gel suspension on a rocker for 1 h at room temperature (or 4 h at 4°C) in a volume sufficient to keep beads mixing.
5. Wash the column with 2× 10 ml of coupling buffer.
6. Block excess active groups with 0.1 M Tris buffer, pH 8, for 1 h at room temperature.
7. Wash away non-covalently bound ligand with 2× 10 ml each of salt buffers of high pH (0.1 M Tris buffer, pH 8, 0.5 M NaCl), followed by low pH (acetic acid, 0.5 M NaCl, pH 4).
8. Equilibrate beads in 3× 10 ml PBS and store at 4°C.

3.3.2. Affinity Purification of Serum Antibodies

1. A 5 ml volume of immunization day 200 cynomolgus monkey anti-RAPTIVA®, hyper-immune serum was diluted in 50 ml PBS.
2. Apply diluted serum to the affinity matrix, dripping by gravity flow, collecting the flow-through and re-applying twice.
3. Wash the column with 3× 10 ml PBS. The antibody-bound column could be stored at 4°C for several days before elution.
4. Elute the bound antibodies in a 10 ml volume by applying 1 ml of elution buffer at a time and collecting into 1-ml Eppendorf tubes containing 10 µL of 1 M Tris, pH 8, to neutralize the acid. Use a one-fourth cut strip of pH paper to check the neutralization, adding more 1 M Tris, pH 8, if necessary to achieve a pH between 5 and 8.
5. Pool the fractions containing protein (determined by protein assay or OD_{280}) and dialyze immediately at 4°C against 2 l of PBS, changed twice.
6. Measure the final protein concentration and store at 4°C for several weeks or longer term at –70°C (see **Note 17**).

3.4. Procedure for the IgGAME Bead Array Assay

3.4.1. Sample and Assay Set-Up

1. Wet the Millipore Multiscreen-BV filter plate wells with 15 µL assay diluent/well.

2. Vortex bead stock solutions and sonicate for 2 min. Add 50 μL per assay well (5000 beads of each bead population) of the IgGAME bead working solution (*see* **Note 11**).

3. Prepare samples at a final concentration of 10% serum. If diluting some serum samples below 10%, dilute them in naive serum to maintain a uniform serum concentration in all samples. Use this same naive serum as a 10% serum background control. Purified samples or controls (serum free) can be assayed in assay diluent, using assay diluent as a background control (*see* **Note 18**).

4. Add 50 μL of samples and controls to each well. Incubate with agitation for 2 h at room temperature. Shorter incubations compromise sensitivity and specificity.

5. Wash the plate 3× 200 μL Phosphate-buffered saline, 0.05% Polysorbate-20 (Tween-20) (PBST) using the vacuum manifold (*see* **Note 19**).

6. Add 100 μL of biotinylated antigen detector/well and incubate 2 h at room temperature. Shorter incubations may reduce sensitivity for slower binding-specific antibodies.

7. Wash the plate 3× 200 μL PBST using the vacuum manifold.

8. Add 160 μL/well of streptavidin-PE. Incubate 30 min at room temperature with agitation (*see* **Note 20**).

9. Add 40 μL/well of 5% formaldehyde (*see* **Note 21**).

10. Measure samples on Luminex® instrument.

3.4.2. Luminex® Instrument Set-Up

1. Set probe at height for filter plates (e.g., on Luminex®-100™ set probe height using three spacer disks in an RIA plate) (*see* **Note 22**).

2. Warm up machine for 30 min. Perform the prime, alcohol flush and wash commands. Check by eye that there is no air in the system tubing and that there is sufficient sheath fluid for the run.

3. Bring the calibrators to room temperature. Calibrate using 5 drops of CAL1 and CAL2 per well in the preset wells of an RIA plate. Daily calibration will ensure the median RFU values are similar from day-to-day between assays. Perform three washes following calibration to flush the calibration beads from the system.

4. Acquire 100 events per well. If instrument allows, set minimum events at 20 (*see* **Note 23**).

5. Set the gate size around 8000–13,500 (*see* **Note 24**).

6. Express data as median RFU. Clean instrument and leave soaking in water.

3.5. Sensitivity Evaluation of IgGAME Assay

1. Purified standards for isotype-specific anti-RAPTIVA® antibodies were not available. Affinity purification of the hyper-immune serum on sepharose covalently linked with RAPTIVA® yielded polyclonal antibody of the IgG class, with very low levels of specific antibodies of the IgA, IgM or IgE classes. A standard curve was generated by spiking known concentrations (10 ng–100 μg/ml) of the purified anti-RAPTIVA® pAb into 10% naive cynomolgus monkey

serum. Data were acquired using singleplex assays with protein G beads or with RAPTIVA®-coupled beads (*see* **Fig. 2**). Approximate sensitivity of the IgA, IgM and IgE components was estimated by capturing purified human myeloma Igs of the appropriate i sotype in assay diluent (no serum) followed by detection with biotinylated anti-human light chain antibodies.

3.6. Competition Studies

1. The specificity of the total antibody response was tested by pre-incubating anti-RAPTIVA® serum at a dilution of 1/10 in assay diluent, with 10 μg/ml of a panel of Genentech recombinant humanized monoclonal IgG antibodies or a commercially available human IgG or F(ab´)$_2$ for 2 h at room temperature. RAPTIVA®-capture beads were then added and the assay performed as usual (*see* **Fig. 4**).

2. The IgE specificity was confirmed using three approaches (*see* **Fig. 5**). First, to check the specificity of capture, 100 μg/ml of anti-IgG, anti-IgA, anti-IgM and anti-IgE monoclonal antibodies were pre-incubated with anti-RAPTIVA® serum at a dilution of 1/10 in assay diluent, overnight at room temperature. Second, to test the antigen specificity, anti-RAPTIVA® serum at a dilution of 1/10 in assay diluent was pre-incubated with 10 μg/ml of several Genentech recombinant humanized monoclonal IgG antibodies or a commercially available human IgG or F(ab´)$_2$ for 2 h at room temperature. Third, capture beads were blocked with 12 μg/ml of human IgG, human IgE or human IgM, overnight at room temperature before adding anti-RAPTIVA® serum.

4. Notes

1. Antibodies conjugated to capture beads should be carefully evaluated for sensitivity and specificity. For example, many anti-human Ig capture antibodies ruled out during the development of this assay were found to cross-react with other Ig isotypes, or be unable to detect the corresponding non-human primate Ig. Several approaches to determine specificity and cross-reactivity are shown in **Figs 1** and **2** and **Tabs 1** and **2**. Also, when testing immunogenicity, the presence of pre-existing serum anti-rodent antibodies binding to the framework regions of the monoclonal antibodies (false-positive results) can be ruled out using pre-immune serum or competition with purified rodent Ig (*see* **Fig. 5**).

2. Anti-human IgM and Protein G Beadlyte™ products from Upstate Inc. may be discontinued. An alternate source of specific anti-IgM antibody characterized to perform well, but with slightly lower sensitivity than the Upstate Inc. product, was clone CM7 from Custom Monoclonals International. Purified protein G for coupling can be purchased from Pierce Biotechnology, Inc. Background due to capture of the IgG therapeutic detector (RAPTIVA®) on protein G was significant, however was largely blocked by the non-specific IgG in serum samples, and could be consistently subtracted. Alternatively, anti-human IgG$_{1,2,3,4}$ subtype antibodies from Zymed Labs (Invitrogen, South San Francisco, CA) performed

well in this system, although no protein of the IgG_3 subtype could be measured in the cynomolgus monkey serum *(5)*.

3. Several commercial Ig light chain detector antibodies that worked well in a standard ELISA format did not work in the Luminex platform. Also, an anti-kappa antibody able to detect all kappa light chains in all cynomolgous monkey isotypes was not identified among commercial vendors. This restriction has limited the use of this technology for quantifying total serum Igs in non-human primates. The light chain detector antibodies used from Upstate Inc. are normally sold as a biotinylated 20× cocktail but can be custom-ordered separately and unlabeled.

4. The optimal coupling concentration for a given protein should be determined. In general for antibody coupling, a ratio of 2.5 µg protein/million beads was found to be optimal (*see* **Fig. 6**). Before coupling, any foreign protein stabilizers (e.g., BSA), azide, glycine, Tris or any other nitrogen-containing chemicals should be removed by dialysis or by using a desalting chromatography column (e.g., Pharmacia Biotech NAP™-10 columns, GE Healthcare Lifesciences, Piscataway, NJ).

5. The carboxylated microspheres should have limited exposure to light. Organic solvents should be avoided, as they will cause the beads to swell and the coding dyes to leach out. Stored in the dark at 4°C, the uncoupled beads have a shelf life of approximately 1 year. When oxidized, the beads will typically shift laterally out of their defined Beadmap zone on the Luminex® instrument. The shelf life of protein-coupled beads can be shorter and was found to vary from 3 to 12 months. When aggregated, the bead counts will be lower than expected and the percent aggregated beads error message during the run will be above 10%.

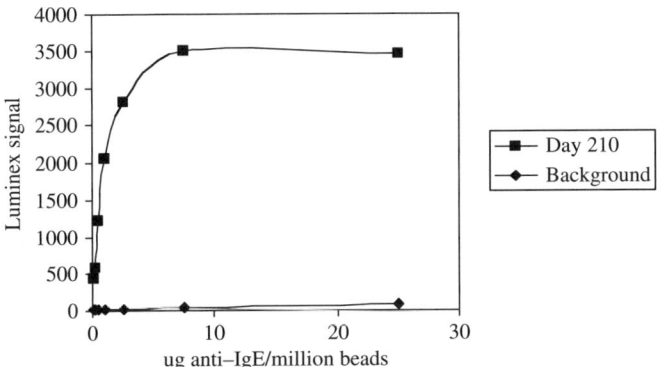

Fig. 6. Optimization of bead coupling. The amount of anti-immunoglobulin E antibody coupled/million microspheres was varied from 0– 25 µg. The sensitivity was tested by measuring the median reactive fluorescent unit signal from day 210 hyper-immunized or naive serum (for background) at a dilution of 1/10 and the biotinylated RAPTIVA®/streptavidin-phycoerythrin detection.

Aggregation appears to be more of an issue once the beads have protein coupled. Most aggregation is reversible by sonicating the stock for 1–5 min.

6. The microspheres have a tendency to stick to each other and to plastic. For long-term storage and especially coupling reactions, the USA Scientific Inc., brand tubes were found to be far superior to other polypropylene or polystyrene brands of tubes, even when the other brands were siliconized. Scaling up of the coupling reaction was done using multiple USA Scientific brand Eppendorf tubes. Storage of beads was found to be best in concentrations in the range of 1×10^7 beads/ml in the dark at 4°C. Working stocks containing 1×10^5 beads/ml for use over several weeks were made from this master stock in 15-mL Blue Max™ Jr. polypropylene falcon tubes (product 352097, Becton Dickenson, Franklin Lakes, NJ) and kept covered in foil at 4°C.

7. The wash steps during coupling, particularly during the bead activation step when the beads form a sheath-like layer, were found to be prone to significant bead losses. Performing the washes of activated beads in a single well of the filter plates improved bead recovery. Manually removing the wash buffer with a fine-tipped transfer pipette is another acceptable alternative.

8. A 1-ml syringe without a needle attachment is useful for measuring and dispensing viscous liquids such as polysorbate-20 (also familiarly called Tween-20).

9. If using the Slide-A-Lyzer® cassettes for dialysis, make sure to inject and then withdraw some air, leaving about 1/10 volume of air for volume expansion during dialysis. If air will not withdraw and the sample does not spread over the area of the cassette membrane, this is will indicate a fatal puncture/loss integrity of the membrane. The sample should be then be transferred to a new cassette immediately. If using tubing for dialysis, tie tight knots on both ends, even if clips are used, again to avoid fatal leaks and loss of precious sample.

10. Activated CH-sepharose 4B is a reactive derivative of sepharose 4B, which provides a 6-carbon spacer arm and an active ester group for spontaneous covalent coupling of proteins and other ligands containing primary amine groups. For linking peptides composed of 30 amino acids of fewer, thiol-activated sepharose beads are recommended. Linkage in this case is through a C- or N-terminal cysteine residue on the peptide and the active thiol groups on the sepharose.

11. There is no advantage towards sensitivity in using more than 5000 beads/well (*see* **Fig. 7**). Using fewer beads increases the total acquisition time per plate and tends to increase the error rate of sampling. It may be useful to pre-test the bead mixture on the Luminex® instrument before adding to the sample wells. As the beads contain internal dyes, there is no need to add a fluorescent tag in order to see them on the instrument (the fluorescent detection step is solely used to recognize and quantify the amount of analyte bound to a bead). Pre-testing the bead preparation will allow any issues with bead numbers, oxidation, aggregation or probe clogs to be evaluated before samples are processed. Add the same volume of beads to a test plate in the same final volume as will be

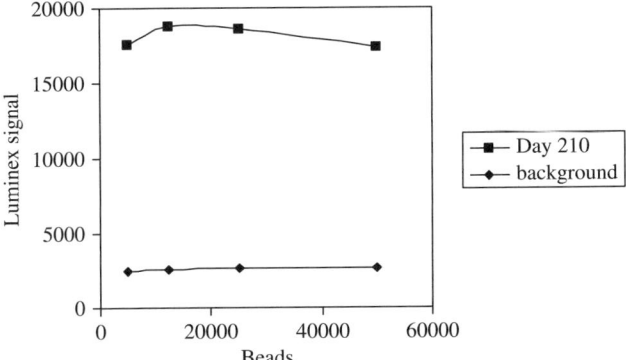

Fig. 7. Optimization of bead input number. The number of beads input into a sample well was varied from 5000 to 50,000. The median reactive fluorescent unit signal from day 210 hyper-immune serum or naive serum (for background), at a dilution of 1/10, using biotinylated RAPTIVA®/streptavidin-phycoerythrin detection, was measured. No significant increase in signal was observed above the 5000 bead input.

used in the assay (in this case 0.2 ml) and run it as a sample. If the probe is clogged, the beads are aggregated or the bead number is lower than expected, there would be very few events collected in the appropriate Beadmap zone. If the beads are oxidized, the bead number will be sufficient, but events will fall outside the target zone and therefore not be counted.

12. The concentration of the detector antibody may vary depending on the sensitivity requirement of the assay for the corresponding analyte and the properties of the detector antibody itself, and should be optimized during assay development. In our case, the 5 µg/ml detector was selected after testing over a range of concentrations from 0.25 to 10 µg/ml.

13. Luminex® instruments and reagents can be purchased directly or from several partners. For all partners supplying instruments, software, service and reagents, check the Luminex® corporate website at www.luminexcorp.com. Any Luminex® instrument will work for this application. Here, the Luminex®-100™ (Luminex Corp.) and BioPlex® (BioRad, Inc., Hercules, CA) were used. Software for data analysis varies with the vendor. In the absence of customized software, data can be directly exported into Excel from the instrument data files. If using BioRad calibrators on instruments other than the BioPlex®, adjust the doublet discriminator value to 10,000. Although the calibration beads are identical to other vendors, the doublet discriminator calibration values marked on the BioRad vials are scaled to fit only with the BioPlex® software.

14. The volume used for the conjugation reaction does not need to be scaled up linearly as long as there is enough volume for mixing.

15. The lot-to-lot variation of the performance of bead conjugates is minimal. However, each new lot should be tested on a control with a known response

in the assay. Unknown sample data responses can be normalized to the control response (or quantified from an intra-assay standard curve) for consistency of data reporting between lots.

16. In some cases, the biotin challenge ratio may need to be optimized for more or less biotin incorporation. Biotin covalently links to lysine residues, and if a particular site is fully occupied, it may affect the ligand-binding activity. Or the efficiency of incorporation may be lower in some molecules, leading to fewer incorporated biotins, resulting in a less-sensitive detection reagent. If desired, the amount of biotin incorporated can be measured by the Avidin-HABA reagent (28010 ImmunoPure® HABA, Pierce Biotechnology, Inc.) or by using a fluorescent biotin for conjugation, such as the FluoReporter® Biotin/DNP (Invitrogen product F-6348, Molecular Probes).

17. Typically, yields of specific antibodies from hyper-immune serum vary from 0.1 to 1 mg/ml serum (the amount depends on the robustness of the immune response to the antigen). After purification, insufficient amounts of anti-RAPTIVA® IgA and IgM and IgE antibodies were present to use as controls. Therefore unpurified serum is used as a source of controls for these isotypes.

18. Naive serum can be selected by screening the baseline response from individuals (if possible the same species as the samples) who have not been treated with the antigen of interest. The final percent serum content of the samples can be lower than 10%, but should be uniform between controls and samples to keep the assay signal proportional to the amount of antigen in the sample. The background signal and the doublet discriminator calibration will change with the buffer composition.

19. Vortexing the plate carefully during washes is optional. After liquid has been drawn through the filter plate, the bottom of the wells should be blotted dry (no banging) to prevent wicking of the liquid in the following steps. Also, the top of the plates should not be tightly sealed with plastic. It is recommended to use the plate's lid and to wrap the plate in foil during incubations.

20. A final volume of no less than 160 μl is recommended. Lower final sample volumes will make acquisitions very sensitive to the precise needle height/dead volume settings. Quantitation of analyte is independent of bead volume.

21. Plate fixation is not necessary unless the total acquisition time for a sample set, normalized to the same controls, exceeds 30 min. The signal and noise in all sample wells will drift upwards proportionally over time until saturation if left unfixed. Controls or standard curves can be used in each plate to ensure appropriate data normalization. Plates can be stored in the dark overnight at 4°C with or without fixation.

22. If the probe height is too high, a sample empty error will occur. If the probe height is too low, the filter plate membrane will be punctured and the well will leak dry. In the event of a leaky well, the beads can be recovered and transferred in assay buffer to a new plate. The final volume of the sample is not important for accuracy as quantitation arises from the amount of median fluorescence

bound on the beads themselves. If possible, have all instrument users agree on a brand of plates and needle height to avoid frequent adjustments.

23. The number of events counted increases statistical significance of the data. In most cases, a bead count per well of 50 events is sufficiently reliable. Some software written for the Luminex® does not allow for a minimum bead count setting. In these cases, it is important to inspect the data and error messages by eye to ensure no samples are reported with values from bead counts less than 50 events. Because data are reported in median RFU, it is possible to get an artificial value if a few carryover beads are counted in a well with no beads. This would occur if there has been a pipetting error in adding beads to a well or a loss of well integrity during the assay. If the probe is cleaned regularly, the instrument has an inherent bead carryover between wells of less than 10%.

24. Although all the Luminex® beads are the same size, the apparent size of the beads changes with the buffer system used. Therefore, the gate settings should be customized to each assay. The purpose of the gate setting is to exclude the measurements of bead aggregates. Any data not acquired from using too narrow a gate setting cannot be recovered.

Acknowledgments

The author would like to thank Dr. David Fei for his advice and encouragement and Cecilia Leddy and Edward O'Hara for technical assistance.

Refernces

1. Wadhwa, M., Bird, C., Dilger, P., Gaines-Das, R. and Thorpe, R. (2003) Strategies for detection, measurement and characterization of unwanted antibodies induced by therapeutic biologicals. Journal of Immunological Methods, **278**, 1–17.

2. Mire-Sluis, A.R., Barrett, Y.C., Devanarayan, V., Koren, E., Liu, H., Maia, M., Parish, T., Scott, G., Shankar, G., Shores, E., Swanson, S.J., Taniguchi, G., Wierda, D. and Zuckerman, L.A. (2004) Recommendations for the design and optimization of immunoassays used in the detection of host antibodies against biotechnology products. Journal of Immunological Methods **289**, 1–16.

3. Quan, F., Matsumoto, T., Shin, Y., Min, Y., Yang, H., Othman, T. and Lee, J. (2004) Relationships between IgG, IgM, IgE resistance to reinfection during the early phase of infection with *Clonorchis sinensis* in rats. Immunological Investigations **33(1)**, 51–60.

4. Biagini, R.E., Moorman, W.J., Lal, J.B., Gallagher, J.S. and Bernstein I.L. (1988) Normal serum IgE and IgG antibody levels in adult male cynomolgus monkeys. Laboratory Animal Science **38(2)**, 194–6.

5. Shearer, M.H., Dark, R.D., Chodosh, J. and Kennedy, R.C. (1999) Comparison and characterization of immunoglobulin G subclasses among primate species. Clinical and Diagnostic Laboratory Immunology **6(6)**, 953–58.

17

Biotin–Protein Bond: Instability and Structural Modification to Provide Stability for In Vivo Applications

Donald M. Mock and Anna Bogusiewicz

Summary

Biotinylation of proteins is a powerful tool for investigating biological phenomenon, both in vitro and in vivo. Biotinylating reagents that form covalent bonds with several types of amino acid residues are commercially available. However, most, if not all, of these commercially available biotinylating agents produce biotin–protein bonds that are susceptible to cleavage in human plasma. Here, we describe the use of immunoglobulin G as a model protein for evaluation of biotin–protein bond stability and for the investigation of the mechanism of biotin release. We also describe the synthesis of a biotin–protein bond that is stable in human plasma and a method for evaluation of that stability.

Key Words: Biotin; biotinylation; avidin; immunoglobulin G; biotinidase; biotinyl-cysteine; biotinyl-cysteinyl-acetyl-IgG.

1. Introduction

Biotinylation of proteins is a powerful tool for investigating biological phenomenon, both in vitro and in vivo. The most widely used in vitro detection systems employ biotinylated antibodies with detection by avidin or streptavidin. Biotinylating reagents that form covalent bonds with several types of amino acid residues are commercially available. However, we have shown that most, if not all, of these commercially available biotinylating agents produce biotin–protein bonds that are susceptible to cleavage in human plasma (*1*). We have also shown that release of biotin from biotinylated proteins occurs both

From: *Methods in Molecular Biology, vol. 418: Avidin-Biotin Interactions, Methods and Applications*
Edited by: R. J. McMahon © Humana Press, Totowa, NJ

enzymatically and nonenzymatically in human plasma *(2)*. Because biotin–protein bond instability in plasma is a major limitation for in vivo studies, we have developed a method that produces a structurally modified biotin–protein bond that is stable in plasma *(3)*.

Here, we describe the use of immunoglobulin G (IgG) as a model protein for evaluation of biotin–protein bond stability and for the investigation of the mechanism of biotin release. We also describe the synthesis of a biotin–protein bond that is stable in human plasma and a method for evaluation of that stability.

2. Materials

2.1. Equipment and Reagents

1. Flat-bottom and u-bottom 96-well plates from Dynex Technologies (Chantilly, VA).
2. Gel filtration 10 DG Econo-Pac columns from Bio-Rad (Hercules, CA).
3. Slide-A-Lyser dialysis cassettes from Pierce (Rockford, IL).
4. Centricon 10-kDa, 30-kDa, 50-kDa, 100-kDa ultrafiltration devices from Millipore (Bedford, MA).
5. Polyvinylidene fluoride (PVDF) membranes from Invitrogen (Carlsbad, CA).
6. Rabbit IgG, dimethyl sulfoxide (DMSO), ethylenediaminetetraacetic acid (EDTA), Tris, polyoxyethylenesorbitan monolaurate (Tween 20), dithiothreitol (DTT), N-morpholinoethane sulfonic acid (MES), Dulbecco's phosphate-buffered saline (PBS), d-biotin, avidin, p-hydroxy-mercuribenzoic acid (HOHgBz), bovine serum albumin (BSA), o-phenylenediamine, and iodoacetic acid N-hydroxysuccinimide ester (iodoacetic acid-NHS), all from Sigma-Aldrich Chemical (St. Louis, MO).
7. Biotin-PEO-amine (BPEO), 5-(biotinamido) pentylamine (5BP), sulfo-NHS-LC-biotin (SNHSB), NHS-LC-biotin (NHSB), iodoacetyl-LC-biotin (IAB), and biotin-LC-hydrazide (BH), NHS-biotin, L-cysteine, 1-ethyl-3-(3-dimethylaminopropyl) carbodiimide HCl (EDC), horseradish peroxidase (HRP)-avidin, BCA reagent, and avidin-alkaline phosphatase from Pierce.
8. N-2-Hydroxyethyl piperazine-N´-2-ethanesulfonic acid (HEPES) [Calbiochem (La Jolla, CA)].
9. gradient Bis-Tris gel, 4–12%, from Invitrogen.
10. ECF™ substrate from Amersham Biosciences (Piscataway, NJ).

2.2. Equipment

1. High performance liquid chromatography (HPLC) system with a binary gradient capability and manual injector with 1-mL sample loop.
2. C18 reverse-phase HPLC column, 2 mm by 25 cm, 0.4 μm particle size.
3. Fraction collector capable of timed advancement.
4. Ninety-six-multiwell plate washer.
5. Multiwell format spectrophotometer with 490-nm and 630-nm wavelength filters.

6. Liquid scintillation counter
7. Gel electrophoresis apparatus and power supply.
8. STORM 840 optical scanner (Molecular Dynamics, Amersham Biosciences, Piscataway, NJ, USA).

3. Methods

This section presents (i) the evaluation of six commercially available biotinylation reagents, (ii) the investigation of the mechanism of instability of the biotin–protein bond, and (iii) the synthesis and characterization of the stability of a biotin–protein bond with an altered structure.

3.1. Selection, Use, and Evaluation of Biotinylation Reagents

For our model protein, we chose IgG purified from rabbit serum. We chose six commercially available biotinylating reagents to evaluate for bond stability. The six actually result in only three chemically distinct links between the biotin moiety and various functional groups on IgG. All reagents share an amide link to the carboxyl group of biotin as do all commercially available biotinylating agents to our knowledge. However, the six differ in the chemical links between the spacer and the various functional groups on IgG.

1. IAB reacts with the SH cysteine residues of IgG.
2. BPEO and 5BP react with carboxyl groups on the protein.
3. NHSB and SNHSB react with ε-amino groups of lysine residues and the N terminus of protein.
4. BH reacts with aldehyde groups produced by oxidized hydroxyl groups on glycoproteins.

3.1.1. Biotinylation of IgG with Different Biotinylation Reagents

3.1.1.1. IODOACETYL-LC-BIOTIN

1. Dissolve 26.7 nmol of IgG in 1 mL of 0.1 M sodium phosphate with 5 mM EDTA (pH 6). Add DTT to a final concentration 50 mM. Vortex for 3 s. Incubate at 37°C for 90 min. Allow the solution to cool at room temperature for 15 min.
2. Use molecular sieve chromatography to remove excess DTT as follows. Apply the reaction mixture onto a gel filtration column (10 DG Econo-Pac) equilibrated with 50 mM Tris with 5 mM EDTA (pH 8.3). Elute with the same buffer.
3. Pool fractions with maximum absorbance at 280 nm.
4. Add 30 μL of 4 mM IAB in DMSO. Incubate mixture for 90 min at room temperature in the dark.
5. Dialyze against 800 mL of 0.1 mM PBS (pH 7.2) for 72 h to remove residual biotinylating reagent and free biotin. Replace dialysis buffer four times over the 72 h (*see* **Note 1**).
6. After dialysis, store sample at –20°C until required.

3.1.1.2. BIOTIN-PEO-AMINE AND 5-(BIOTINAMIDO) PENTYLAMINE

1. Dissolve 16.7 nmol of IgG in 0.5 mL of 0.1 M MES buffer (pH 5.5). Add 25 µL of an aqueous solution of 50 mM BPEO or 5BP. Mix and add 6.25 µL of 5 mM EDC in 0.1 M MES. Incubate for 2 h at room temperature.
2. Dialyze against 800 mL of 0.1 mM PBS (pH 7.2) for 72 h.

3.1.1.3. SULFO-NHS-LC-BIOTIN

1. Dissolve 13.3 nmol of IgG in 1 mL of 0.1 M PBS (pH 7.2). Add 20 µL of an aqueous solution of 18 mM SNHSB. Incubate for 2 h on ice.
2. Dialyze against 800 mL of 0.1 mM PBS (pH 7.2) for 72 h.

3.1.1.4. NHS-LC-BIOTIN

1. Dissolve 13.3 nmol of IgG in 1 mL of 0.1 M PBS buffer (pH 7.2). Add 100 µL of 182.2 mM NHSB in DMSO. Incubate for 2 h on ice.
2. Dialyze against 800 mL of 0.1 mM PBS (pH 7.2) for 72 h.

3.1.1.5. BIOTIN-LC-HYDRAZIDE

1. Incubate 133 nmol of IgG in 0.1 mM acetate with 0.02% Tween 20 and 10 mM sodium periodate (pH 5.5) at 4°C for 30 min.
2. To remove periodate, load this mixture onto a gel filtration column (10 DG Econo-Pac) equilibrated with acetate buffer; elute with the same buffer.
3. Pool the two 1 mL fractions with maximum absorbance at 280 nm.
4. Add 0.5 mL of an aqueous solution of 20 mM BH. Incubate for 2 h at room temperature.
5. To separate biotinylated protein from residual biotinylation reagent and low molecular weight hydrolysis products, load this mixture onto a gel filtration column (10 DG Econo-Pac) equilibrated with PBS with 0.02% Tween buffer. Elute with the same buffer. Collect and pool the two 1 mL fractions with the maximum absorbance at 280 nm.

3.1.2. Determination of Stoichiometry of Biotinylation

Total biotin covalently bound to IgG was determined as free biotin released by acid hydrolysis. Released biotin was determined by an avidin-binding assay as described previously *(4,5)*. Briefly, biotin bound to IgG was hydrolyzed by incubation with 1.5 M HCl at 100°C for 120 min. Hydrolysates were analyzed after a dilution of ≥100-fold in assay buffer (0.1 M HEPES, 1 M NaCl, and 0.1% Tween 20 at pH 7.0) and thus did not require HPLC separation to remove interfering substances (*see* **Note 2**). Diluted hydrolysates were incubated with avidin coupled to HRP (HRP–avidin) and then transferred to a microtiter well coated with biotinylated BSA. Those HRP–avidin molecules with unoccupied biotin-binding sites will bind to the biotin moieties on the immobilized BSA. The more released biotin, the fewer HRP–avidin molecules are available to bind to the biotinyl-BSA. The amount of bound avidin is quantitated by bound HRP using optical detection of the rate of oxidation of *O*-phenylenediamine. Optical density at 490 nm was determined using a multiwell spectrophotometer with background correction at 630 nm. The absolute concentration of biotin was determined by comparison to a standard curve constructed from concentrations of biotin determined gravimetrically and confirmed radiometrically from ^{3}H-biotin (*see* **Note 3**). Protein concentrations were determined by a BCA protein assay (Pierce). The stoichiometry, calculated as the molar ratio of total biotin to IgG, varied among biotinylating reagents (*see* **Table 1**). Maximal protein biotinylation was obtained using either NHSB or SNHSB.

3.1.3. Quantitation of the Biotin–IgG Stability

Evaluation of the biotin–IgG stability is based on the quantitation of biotin release from IgG in the presence of either PBS (control) or human plasma. A critical aspect of quantitating biotin release by human plasma is removal of free biotin from the plasma prior to the use in the stability studies (*see* **Note 4**).

1. Dialyze human plasma against PBS buffer containing avidin with a calculated biotin-binding capacity equal to 2.5 times the estimated biotin content of the plasma.
2. Add 10 µL of biotinylated IgG to 990 µL of human debiotinylated plasma from which biotin has been removed and to PBS buffer. Vortex gently and incubate for 4 h at room temperature.
3. Dilute mixture with water 1:1 (volume: volume) and separate released biotin from biotinylated IgG by ultrafiltration using a Centricon 10-kDa MW cut-off unit (*see* **Note 5**).

Table 1
Stoichiometry of Biotinylation of Immunoglobulin G (IgG)

Biotinylation reagent	Biotin/protein (mol/mol)
Biotin-PEO-amine	0.3
5-(Biotinamido)-pentylamine	0.2
Iodoacetyl-LC-biotin	0.4
Biotin-LC-hydrazide	2.7
Sulfo-NHS-LC-biotin	4.3
NHS-LC-biotin	4.5
Biotinyl-cysteine	3.6

Degree of biotinylation was determined by the horseradish peroxidase–avidin-binding assay of biotinylated IgG after acid hydrolysis to release covalently bound biotin.

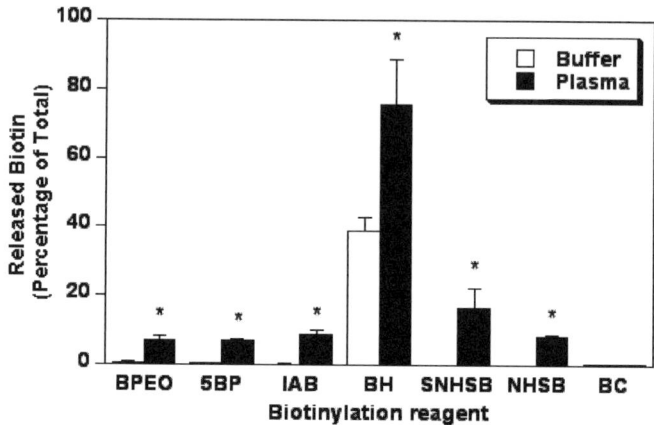

Fig. 1. Biotin released expressed as percentage of total biotin. Biotin was quantitated after 4-h incubation with phosphate-buffered saline buffer (control) or human plasma. Released biotin was detected by horseradish peroxidase–avidin-binding assay of the 10-kDa ultrafiltrate. The bond formed by using biotinyl-cysteine (BC) is stable in plasma, but other bonds are not. *Significantly different ($P < 0.0001$) from control. BC, biotinyl-cysteine; BH, biotin-LC-hydrazide; 5BP, 5-(biotinamido) pentylamine; BPEO, biotin-PEO-amine; IAB, iodoacetyl-LC-biotin; NHSB, NHS-LC-biotin; SNHSB, sulfo-NHS-LC-biotin.

4. Assay ultrafiltrate for biotin concentration using an avidin-binding assay as previously described *(5)*.

Biotin released by plasma or in the buffer control is expressed as percentage of total biotin determined by acid hydrolysis (*see* **Fig. 1**). Five of the six biotinylating reagents examined produced bonds that were stable in buffer ($\leq 0.6\%$ loss of total biotin). In contrast for BH, 39% of the total biotin label was released by 4-h incubation with buffer. Incubation with plasma significantly increased ($P < 0.0001$) biotin release for all six reagents. For the five labels stable in buffer, the average increase in total biotin released was 26-fold.

3.2. Investigation of the Mechanism of Biotin Release from Biotinylated Proteins in Plasma

Described below is an approach for investigation of the mechanism of biotin release from biotinylated proteins in plasma. Included are methods for plasma preparation, incubation of biotinylated proteins with plasma or plasma treated to either inactivate enzymes or remove all macromolecules, separation of released biotin by ultrafiltration, quantitation of released biotin by the

HRP–avidin-binding assay, HPLC determination of the chemical nature of released biotin moieties, and an examination of protein integrity.

3.2.1. Preparation of Plasma

3.2.1.1. INHIBITION OF PLASMA BIOTINIDASE

1. Incubate biotin-free plasma at room temperature for 10 min with an aqueous solution of 0.1 mM HOHgBz.
2. Use immediately.

3.2.1.2. REMOVAL OF MOLECULES WITH MOLECULAR WEIGHT GREATER THAN APPROXIMATELY 10 KDA BY ULTRAFILTRATION

1. Ultrafilter biotin-free plasma by centrifugation through Centricon 100-kDa cut-off membrane at 1000 g for 30 min.
2. Ultrafilter the ultrafiltrate from the first separation sequentially through Centricon 50-, 30-, and 10-kDa cut-off membranes at 5000 g and for 30 min, 30 min, and 1 h, respectively

3.2.1.3. HEAT DENATURATION

1. Incubate biotin-free plasma at 100°C for 15 min.

In previous studies, inactivation of enzymes or elimination of all macro-molecules with molecular weight >10 kDa reduced the rate of biotin release by about 60%. We concluded that biotin was being released by both enzymatic and non-enzymatic mechanisms *(2)*.

3.2.2. Quantitation of Released Biotin

1. Add 10 µL of biotinylated IgG to 990 µL of biotin-free human plasma, HOHgBz-treated plasma, ultrafiltrate of plasma, heat-treated plasma, or PBS buffer. Vortex gently and incubate for 4 h at room temperature.
2. Dilute mixtures with water 1:1 (volume : volume).
3. Separate released biotin from biotinylated IgG by ultrafiltration using a Centricon 10-kDa MW cut-off membrane.
4. Assay the ultrafiltrate for biotin using the HRP–avidin-binding assay as previously described *(5)*.

3.2.3. HPLC Characterization of Released Biotin Moieties

The biotin bond to IgG may be disrupted in several places. If the biotin bond at the amide bond nearest biotin is hydrolyzed, the moiety released by hydrolysis would be free biotin and thus will have the same HPLC retention time as authentic biotin. If hydrolysis occurs at the bond nearest to the protein, the released moiety will contain biotin, but will retain the spacer arm; this moiety should elute at a retention time different from that of biotin. Described below are the steps that can be utilized in identification of the released biotin moiety.

1. Add 10 µL of biotinylated IgG to 990 µL of biotin-free human plasma. Vortex gently and incubate for 4 h at room temperature.
2. Dilute mixtures with water 1:1 (volume : volume) and separate released biotin from biotinylated IgG by ultrafiltration using a Centricon 10-kDa MW cut-off unit.
3. Equilibrate a C18 reverse-phase column in 0.05% (v/v) trifluorocetic acid (TFA) adjusted to pH 2.5 with ammonium acetate (solution A). The HPLC flow rate is constant at 1 mL/min. Inject 200 µL of ultrafiltrate. Elute with a linear 0–40% gradient of acetonitrile: 0.05% TFA (1:1, v/v) (solution B) over 35 min.
4. Collect 1 mL fractions. Neutralize each with 120 µL of a 1.5:1 mixture of 0.125 M NaOH : 0.02 M HEPES buffer.
5. Dry the fractions to remove the chromatography solvents.
6. Dissolve the dried samples with 0.5 mL of deionized water for 30 min before adding 0.5 mL of HEPES assay buffer.
7. Assay 1-mL samples for avidin-binding substances as previously described *(5)*.
8. Compare the retention time of any avidin-binding activity in the fractions to the retention time of [^{3}H] biotin previously determined.

Based on a previous study *(2)*, the released moiety is free biotin. This finding suggested that the link between biotin and the spacer arm is being hydrolyzed by some factor in plasma.

3.2.4. Confirmation of Protein Integrity

The steps described here outline an approach to evaluating the integrity of the biotinylated model protein (IgG). If low-molecular-mass biotinylated protein fragments were released into the ultrafiltrate, this would mandate an alternative interpretation of the site of bond instability, the stoichiometry of biotinylation, and mechanism of instability as described above.

1. Add 10 µL of biotinylated IgG to 990 µL of human plasma. Vortex gently and incubate for 4 h at room temperature.
2. Dilute mixture with water 1:1 (volume : volume). Load into a Centricon 100-kDa MW cut-off membrane and ultrafilter at 1000 g for 30 min.
3. Dissolve 20 µL of ultrafiltrate samples in 5 µL of sodium dodecyl sulfate. Heat at 90°C for 10 min, electrophorese and blot as follows.
4. Load on a pre-cast, commercially available 4–12% gradient Bis-Tris gel (Invitrogen) (*see* **Note 6**).
5. Run gel at 200 V constant for 50 min at starting current of 115 mA.
6. Electroblot proteins to a PVDF for 1 h at 30 V constant at starting current of 170 mA.
7. Block membrane with 0.05% Tween 20 in PBS for 1 h at room temperature while shaking.

8. Incubate membrane with 1 µg/mL avidin-alkaline phosphatase (Pierce) in 50 mL blocking buffer for 1 h at room temperature while shaking.
9. Wash membrane three times in blocking buffer for 5 min at room temperature while shaking.
10. Identify the bands by incubation for 5 min with 1 mL of ECF™ substrate (Amersham Biosciences) in a sheet protector at room temperature (*see* **Note 7**).
11. Quantitate fluorescence of bands using a STORM 840 optical scanner (*see* **Note 8**).
12. Estimate relative intensity of bands by Image Quant software program (Molecular Dynamics).

If this procedure demonstrates no degradation of the biotinylated protein, then the previous interpretations of bond stability and stoichiometry are valid. In our studies *(1,2)*, the IgG portion of biotinylated protein was intact.

3.3. Synthesis and Characterization of Biotinyl-Cysteinyl-Acetyl-IgG

The synthetic scheme for the preparation of biotinyl-cysteinyl-acetyl-IgG (Bio-C-A-IgG) (*see* **Fig. 2**) shows the final linkage between biotin and the ε-amino groups of the lysine residues in IgG contains a carboxyl group attached to the carbon that is alpha to the biotinamide bond. The synthesis of Bio-C-A-IgG required the formation of two intermediate compounds: biotinyl-cysteine (BC) and iodo-acetyl-IgG (IA-IgG). The synthesis is described in **Subheadings 3.3.1.** and **3.3.2.**

3.3.1. Biotinyl-Cysteine

1. Mix 1 mL of a L-cysteine solution (37.6 µmol) in 0.1 M NaH_2PO_4 with 1 mL of an NHS-biotin solution (18.7 µmol) in 70% methanol in water (18.7 µmol). Adjust the pH of the mixture to 8.0. Agitate at room temperature for 1 h.
2. Adjust the reaction mixture to pH 2.0 with HCl. Incubate overnight at 4°C. BC crystals will form.
3. Wash the crystals three times with water. Dissolve the crystals in 500 µL of 70% methanol in 0.1 M NaH_2PO_4.

3.3.2. Iodo-Acetyl-IgG

1. Mix 1 mL of a solution of IgG (13.4 nmol) in PBS with 76 µL of an iodoacetic acid-NHS (270 nmol) in DMSO. Shake at room temperature for 1 h.
2. Add approximately 6 µL of Tween 20 diluted 100-fold in PBS to produce a final concentration of 0.05 mM Tween 20.
3. Separate IA-IgG from lower molecular weight reagents and products by dialysis against 800 mL of PBS with 0.02% Tween 20 using an Amicon 10-kDa cut-off Slide-A-lyser dialysis cassette.

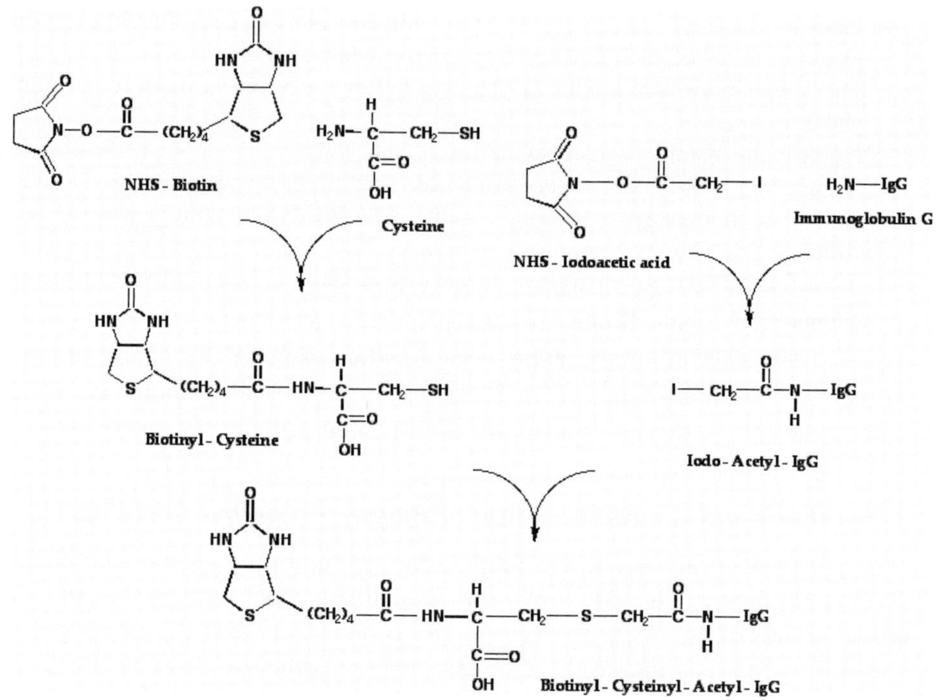

Fig. 2. Reaction scheme for the synthesis of biotinyl-cysteinyl-acetyl-IgG.

3.3.3. Biotinyl-Cysteinyl-Acetyl-IgG

1. Mix a dialyzed solution of IA-IgG (~1.3 ml) with 100 µL of BC in 70% methanol in 0.1 M NaH_2PO_4. Adjust the pH of the reaction mixture to 8.2.
2. Shake at room temperature overnight.
3. Dialyze against 800 mL PBS with 0.02% Tween 20.

Determination of the stoichiometry of biotinylation and quantitation of stability for Bio-C-A-IgG were performed as described in **Subheadings 3.1.2.** and **3.1.3.** The stoichiometry (mol of biotin/mol of IgG) was 3.6 (*see* **Table 1**); this stoichiometry is similar to that achieved by other biotinylating agents. The new synthetic scheme produced Bio-C-A-IgG that was stable in both buffer and plasma. Only 0.53% of the total biotin label was released by incubation in plasma (*see* **Fig. 1**); this rate is not significantly different from control. This finding provides evidence that the instability of the biotin bond of the other biotinylating agents does indeed originate from the amide bond to the spacer arm, not the bond to the amino acid residue.

4. Notes

1. For dialysis of small volumes (1–2 mL,) Slide-A-Lyser cassettes (Pierce) are useful. Purification of biotinylated protein requires at least 3 days of extensive dialysis with a change of buffer every 12 h.
2. When using undiluted acid hydrolysates, the major concern is the prevention of artifactual increases in the apparent biotin resulting from substances produced during acid hydrolysis as documented previously (*4,5*).
3. All samples were assayed at least in triplicate against the biotin standard curve constructed by diluting a known concentration of biotin. Biotin (d-biotin) was obtained from Sigma. Biotin stock was prepared as previously described by Mock (*5*). Assay standards were prepared by diluting a 100 nM biotin solution to a final concentration of 3000 pM. Biotin standard concentrations were confirmed by comparison with [^{3}H]-biotin of a known specific activity (Perkin Elmer Life and Analytical Sciences, Inc. Boston, MA).
4. The normal range of biotin content of the plasma is 294–1021 pM (*6*). After dialysis, the free biotin concentration in the plasma was undetectable ($\leq$22 pM).
5. An initial 1:1 dilution with water prevents an artifactual decrease in the concentration of free biotin passing through the membrane, presumably by preventing a build-up of high molecular weight complexes on the Centricon membrane.
6. Gel run as per manufacturer's instructions.
7. Be careful not to disturb membrane during incubation with the ECF substrate; after 5-min incubation, carefully use forceps to lift membrane from the substrate. Allow excess liquid to drain off and place in new clean sheet protector for reading on the STORM instrument.
8. This method allows detection of as little as 2 pg of biotinylated BSA.

Acknowledgments

This work was supported by PO1 HL 46925 of the National Institute of Heart, Blood, and Lung and by RO1 DK 36823 of the National Institutes of Diabetes, Digestive, and Kidney Diseases.

References

1. Bogusiewicz, A., Mock, N. I., and Mock, D. M. (2004). Instability of the biotin-protein bond in human plasma. *Anal. Biochem.* **327**, 156–161.
2. Bogusiewicz, A., Mock, N. I., and Mock, D. M. (2004). Release of biotin from biotinylated proteins occurs enzymatically and nonenzymatically in human plasma. *Anal. Biochem.* **331**, 260–266.
3. Bogusiewicz, A., Mock, N. I., and Mock, D. M. (2004). A biotin-protein bond with stability in plasma. *Anal. Biochem.* **337**, 98–102.
4. Mock, D. M. and Malik, M. I. (1992). Distribution of biotin in human plasma: most of the biotin is not bound to protein. *Am. J. Clin. Nutr.* **56**, 427–432.

5. Mock, D. M. (1997). Determinations of biotin in biological fluids. In *Methods in Enzymology* (McCormick, D. B., Suttie, J. W. and Wagner, C., eds.), Academic Press, New York, NY. **279**, 265–275.

6. Mock, D. M., Lankford, G. L., and Mock, N. I. (1995). Biotin accounts for only half of the total avidin-binding substances in human serum. *J. Nutr.* **125**, 941–946.

Index